Basic Concepts in Clinical Biochemistry:
A Practical Guide

Vijay Kumar • Kiran Dip Gill

Basic Concepts in Clinical Biochemistry: A Practical Guide

Vijay Kumar
Department of Biochemistry
Maharshi Dayanand University
Rohtak, Haryana, India

Kiran Dip Gill
Department of Biochemistry
Postgraduate Institute of Medical Education
& Research
Chandigarh, India

ISBN 978-981-10-8185-9   ISBN 978-981-10-8186-6   (eBook)
https://doi.org/10.1007/978-981-10-8186-6

Library of Congress Control Number: 2018935904

© Springer Nature Singapore Pte Ltd. 2018
This work is subject to copyright. All rights are reserved by the Publisher, whether the whole or part of the material is concerned, specifically the rights of translation, reprinting, reuse of illustrations, recitation, broadcasting, reproduction on microfilms or in any other physical way, and transmission or information storage and retrieval, electronic adaptation, computer software, or by similar or dissimilar methodology now known or hereafter developed.
The use of general descriptive names, registered names, trademarks, service marks, etc. in this publication does not imply, even in the absence of a specific statement, that such names are exempt from the relevant protective laws and regulations and therefore free for general use.
The publisher, the authors and the editors are safe to assume that the advice and information in this book are believed to be true and accurate at the date of publication. Neither the publisher nor the authors or the editors give a warranty, express or implied, with respect to the material contained herein or for any errors or omissions that may have been made. The publisher remains neutral with regard to jurisdictional claims in published maps and institutional affiliations.

Printed on acid-free paper

This Springer imprint is published by the registered company Springer Nature Singapore Pte Ltd.
The registered company address is: 152 Beach Road, #21-01/04 Gateway East, Singapore 189721, Singapore

Preface

Biochemistry is one of the fundamental subjects of life science, and knowledge of its practical aspects is absolutely essential for students and researchers. Performing a practical in biochemistry requires adequate facilities and training laboratories for organizing meaningful practicals both at undergraduate and postgraduate levels. Clinical biochemistry is a diagnostics subject, which aims to use standard methods, to monitor disease development and treatment by biochemical methods. Laboratory investigations pertaining to clinical biochemistry provide useful information to clinicians, both in the diagnosis of illness and the monitoring of treatment. The necessity for a practical manual on orientation toward various aspects of clinical biochemistry is always desired. The present manual is written in simple language covering various experimental aspects related with clinical biochemistry such as urinary analysis of bioorganic constituents and estimation of proteins, urea, sugar, uric acid, bilirubin, etc., in blood which is essential for early diagnosis of a disease and also assessment of its therapy efficacy.

The book is specifically helpful to BSc, MLT, MSc, and MSc biochemistry students. Each chapter begins with theoretical aspects of the practical, and the experiments are further well supported by self-explanatory tables and simple calculations that will make the students understand protocols very easily. The manual is also helpful to teachers who are taking clinical biochemistry practicals to guide students in a simple way.

The authors hope that the manual will meet the requirement of undergraduate and postgraduate students studying biochemistry and will be glad to accept constructive criticisms and suggestions from the faculty, students, and readers to make this manual a better one in the future.

Rohtak, Haryana, India	Vijay Kumar
Chandigarh, India	Kiran Dip Gill

Reference Biochemical Values in Serum/Plasma

Acid phosphatase	1–4 KAU/dl	Albumin	3.5–5 g/dl
Total proteins	3.5–5.5 g/dl	Globulins	2.5–3.5 g/dl
Alaninine transaminase (ALT, SGPT)	3–40 IU/L	Alkaline phosphatase	3–13 KAU/dl
		Amylase	5–45 SU/dl
Aspartate transaminase (AST, SGOT)	5–45 IU/L	Calcium	9–11 mg/dl
		Calcium (U)	100–200 mg/day
Bilirubin (total)	0.2–1 mg/dl	Ceruloplasmin	25–50 mg/dl
Bilirubin direct	< 0.4 mg/dl	Cholesterol	150–200 mg/dl
Bilirubin indirect	< 0.6 mg/dl	HDL	30–60 mg/dl
Creatinine	0.5–1.5 mg/dl	LDL	60–130 mg/dl
Creatinine (U)	0.8–2 g/day	VLDL	20–40 mg/dl
Chloride	103–105 mEq/L	Inorganic phosphorus	3.0–4.5 mg/dl
Creatine kinase, CK-MB	24–195 U/L	Phosphorus (U)	0.5–1.5 g/day
Creatine kinase, total	10–400 U/L	LDH	100–190 U/L
Fibrinogen	0.2–0.4 g/dl	Urea	15–40 mg/dl
Glucose, fasting	70–100 mg/dl	Urea (U)	15–40 g/day
Lipase	0.5–160 U/L	Uric acid	3.5–8 mg/dl
Transferrin	200–300 mg/dl	Uric acid (U)	200–500 mg/day
Triglyceride, fasting	50–175 mg/dl	Urobilinogen (U)	1–3.5 mg/day
Sodium	135–145 mEq/L		
Potassium	3.5–5 mEq/L		
Chloride	103–105 mEq/L		

U -Urine

Contents

1	**Common Clinical Laboratory Hazards and Waste Disposal**	1
	1.1 Waste Disposal in Laboratory	2
2	**Blood Collection and Preservation**	5
	2.1 Blood Collection	5
3	**Quality Control in Laboratory**	9
	3.1 Types of Laboratory Errors	10
	3.2 Methods to Minimize the Laboratory Errors	11
4	**Automation in Clinical Laboratory**	13
	4.1 Types of Autoanalyzers	13
5	**Photometry: Colorimeter and Spectrophotometer**	17
	5.1 Colorimeter and Spectrophotometer	18
6	**Preparation of General Laboratory Solutions and Buffers**	21
	6.1 Molar Solutions	21
	6.1.1 Molarity (M)	21
	6.2 Normal Solutions	22
	6.2.1 Normality (N)	23
	6.3 Percent (%) Solutions	24
	6.4 Buffer Solutions	24
	6.5 Physiological Buffers in the Human Body	25
	6.6 Preparation of Common Laboratory Buffers	25
	6.6.1 0.2 M Acetate Buffer (pKa 4.86)	25
	6.6.2 0.2 M Sodium Phosphate Buffer (pKa 6.86)	26
	6.6.3 0.2 M Tris-HCl Buffer (pKa 8.1)	26
7	**Examination of Urine for Normal Constituents**	29
	7.1 Preservatives Used for Urine Collection	29
	7.2 Physical Examination of Urine	30
	7.2.1 Color and Odor	30
	7.2.2 Appearance	30

		7.2.3	Specific Gravity	30
		7.2.4	Volume	30
		7.2.5	pH	31
	7.3	Chemical Examination of Urine		31
	7.4	Tests for Inorganic Constituents of Urine		31
	7.5	Tests for Organic Constituents		32
	7.6	Detection of Urea		32

8 To Perform Qualitative Tests for Urinary Proteins 33
    8.1   Theory 33
    8.2   Tests for Urinary Proteins 34
          8.2.1   Dipstick Test 34
          8.2.2   The Boiling Test for Coagulable Proteins 34
          8.2.3   Sulphosalicylic Acid Test 35
          8.2.4   Nitric Acid Ring Test (Heller's Test) 35
          8.2.5   Bence-Jones Proteins 35

9 To Determine the Quantity of Proteins in Urine Sample Using Biuret Reaction 39
    9.1   Theory 39
    9.2   Specimen Requirements 39
    9.3   Principle 39
    9.4   Reagents 40
    9.5   Calculations 41
    9.6   Clinical Significance 41

10 To Estimate the Amount of Total Protein and Albumin in Serum and to Find A/G Ratio 43
    10.1   Theory 43
    10.2   Principle 43
    10.3   Specimen Requirements 44
    10.4   Reagent 44
    10.5   Procedure 45
    10.6   Calculations 45
    10.7   Estimation of Albumin in Serum 46
    10.8   Principle 46
    10.9   Reagents 46
    10.10   Procedure 46
    10.11   Calculations 47
    10.12   Clinical Significance 47
    10.13   Precautions 48

11	**To Perform Qualitative Test for Reducing Substances in Urine**		**49**
	11.1	Theory	49
	11.2	Qualitative Test for Reducing Sugars	49
		11.2.1 Benedict's Test	49
		11.2.2 Benedict's Qualitative Reagent	50
		11.2.3 Procedure	50
	11.3	CLINISTIX/Uristix	50
		11.3.1 Procedure	51
	11.4	Precautions	51
	11.5	Clinical Significance	51
12	**Quantitative Analysis of Reducing Sugars in Urine**		**53**
	12.1	Theory	53
	12.2	Specimen Requirement	53
	12.3	Principle	53
	12.4	Reaction	54
	12.5	Reagents	54
	12.6	Procedure	54
	12.7	Precautions	54
	12.8	Calculation	55
	12.9	Clinical Significance	55
13	**Estimation of Blood Glucose Levels by Glucose Oxidase Method**		**57**
	13.1	Theory	57
	13.2	Principle	57
	13.3	Specimen Requirements	58
	13.4	Reagents	58
	13.5	Procedure	58
	13.6	Calculations	59
	13.7	Clinical Significance	59
14	**To Determine the Blood Glucose Levels by Folin and Wu Method**		**61**
	14.1	Principle	61
	14.2	Reagents	61
	14.3	Procedure	62
	14.4	Calculations	62
15	**To Perform Glucose Tolerance Test**		**63**
	15.1	Method of Carrying GTT	63
	15.2	Specimen Requirements	63
	15.3	Principle	63
	15.4	Reagents	64
	15.5	Procedure	64
	15.6	Calculations	66
	15.7	Clinical Significance	66

16	**Estimation of Urea in Serum and Urine**	67
	16.1 Theory	67
	16.2 Specimen Requirements	67
	16.3 Principle	68
	16.4 Reagents	68
	16.5 Procedure	69
	16.6 Calculations	69
	16.7 Clinical Significance	70
17	**To Determine Urea Clearance**	71
	17.1 Theory	71
	17.2 Maximum Urea Clearance	71
	17.3 Standard Urea Clearance	72
	17.4 Procedure	72
	17.5 Precautions	72
	17.6 Clinical Significance	72
18	**To Estimate Creatinine Level in Serum and Urine by Jaffe's Reaction**	75
	18.1 Theory	75
	18.2 Specimen Requirements	75
	18.3 Principle	76
	18.4 Reagents	77
	18.5 Preparation of Protein-Free Filtrate	77
	18.6 Procedure	77
	18.7 Calculations	78
	18.8 Clinical Significance	78
19	**To Determine Creatinine Clearance**	79
	19.1 Theory	79
	19.2 Specimen Requirement and Procedure	79
	19.3 Clinical Significance	80
20	**To Determine the Uric Acid Concentration in Serum and Urine**	81
	20.1 Theory	81
	20.2 Specimen Requirements	81
	20.3 Principle	82
	20.4 Reagents	82
	20.5 Procedure	82
	20.6 Calculations	83
	20.7 Clinical Significance	83
	20.8 Precautions	84
21	**Estimation of Total Calcium in Serum and Urine**	85
	21.1 Theory	85
	21.2 Specimen Requirements	85

21.3	Method	86
21.4	Principle	86
21.5	Reagents	86
21.6	Procedure	86
21.7	Calculations	87
21.8	Clinical Significance	87

22 Estimation of Inorganic Phosphorus in Serum and Urine 89

22.1	Theory	89
22.2	Specimen Requirements	90
22.3	Methodology	90
22.4	Principle	90
22.5	Reagents	90
22.6	Preparation of Protein-Free Filtrates	90
22.7	Procedure	91
22.8	Calculations	91
22.9	Clinical Significance	92
22.10	Precautions	92

23 To Estimate the Amount of Total Cholesterol in Serum 93

23.1	Theory	93
23.2	Specimen Requirements	94
23.3	Methodology	94
23.4	Principle	94
23.5	Reagents	94
23.6	Procedure	94
23.7	Calculations	95
23.8	Clinical Significance	95

24 To Estimate Total and Direct Bilirubin in Serum 97

24.1	Theory	97
24.2	Specimen Requirements	97
24.3	Principle	98
24.4	Reagents	99
24.5	Procedure	99
24.6	Calculations	100
24.7	Clinical Significance	100

25 To Determine Alanine and Aspartate Transaminase Activity in Serum 103

25.1	Theory	103
25.2	Specimen Requirements	103
25.3	Principle	103
25.4	Reagents	104
25.5	Procedure	105
25.6	Calculation	106
25.7	Clinical Significance	106

26	**To Estimate the Activity of Alkaline Phosphatase in Serum**	107
	26.1 Theory	107
	26.2 Specimen Requirements	107
	26.3 Principle	107
	26.4 Reagents	108
	26.5 Procedure	108
	26.6 Calculations	109
	26.7 Clinical Significance	109
27	**To Estimate the Activity of Acid Phosphatase in Serum**	111
	27.1 Theory	111
	27.2 Specimen Requirements	111
	27.3 Principle	111
	27.4 Reagents	111
	27.5 Procedure	112
	27.6 Calculations	112
	27.7 Clinical Significance	112
28	**To Determine Serum and Urinary Amylase Activity**	113
	28.1 Theory	113
	28.2 Specimen Requirements	113
	28.3 Principle	113
	28.4 Reagents	114
	28.5 Procedure	114
	28.6 Calculation	114
	28.7 Clinical Significance	115
29	**To Estimate the Activity of Lipase in Serum**	117
	29.1 Theory	117
	29.2 Specimen Requirements	117
	29.3 Principle	117
	29.4 Reagents	118
	29.5 Procedure	118
	29.6 Calculation	118
	29.7 Clinical Significance	118
30	**Qualitative Analysis of Ketone Bodies in Urine**	119
	30.1 Theory	119
	30.2 Rothera's Test for Acetoacetic Acid and Acetone	120
	30.2.1 Principle	120
	30.2.2 Reagents	120
	30.2.3 Procedure	120
	30.2.4 Result	121
	30.3 Gerhardt's $FeCl_3$ Test for Acetoacetic Acid	121
	30.3.1 Reagent	121
	30.3.2 Procedure	121

	30.4	Ketostix Test for Acetone and Acetoacetate	121
		30.4.1 Procedure	121
	30.5	Detection of β-Hydroxybutyrate	122
		30.5.1 Principle	122
		30.5.2 Procedure	122
	30.6	Clinical Significance	122
31	**Qualitative Test for Bile Pigments and Urobilinogen in Urine**		123
	31.1	Theory	123
	31.2	Tests for Bile Pigments in Urine	123
		31.2.1 Fouchet's Test	123
		31.2.2 Hunter's Test	125
		31.2.3 Gmelin's Test	125
	31.3	Urobilinogen in Urine	126
		31.3.1 Ehrlich's Test	126
		31.3.2 Saturated Sodium Acetate Solution	126
	31.4	Precautions	126
	31.5	Clinical Significance	127
32	**Determination of Total Lactate Dehydrogenase Activity in Serum Sample**		129
	32.1	Theory	129
	32.2	Specimen Requirements	129
	32.3	Principle	129
	32.4	Reagents	130
	32.5	Procedure	130
	32.6	Calculation	130
	32.7	Clinical Significance	130
33	**To Measure Activity of Creatine Kinase in Serum**		131
	33.1	Theory	131
	33.2	Sample Requirement	131
	33.3	Principle	131
	33.4	Enzyme Reagents	132
	33.5	Procedure	132
	33.6	Calculation	132
	33.7	Clinical Significance	133
34	**Analysis of Cerebrospinal Fluid for Proteins and Sugars**		135
	34.1	Theory	135
	34.2	Analysis of Proteins in CSF	135
		34.2.1 Pyrogallol Dye-Binding Method	135
		34.2.2 Turbidimetry Method	136
	34.3	Analysis of CSF Glucose	137
	34.4	Clinical Significance	137

35 To Measure Lipid Profile in Serum Sample 139
- 35.1 Sample Requirement 139
- 35.2 Total Cholesterol Estimation 139
  - 35.2.1 Principle 139
  - 35.2.2 Reagents 140
  - 35.2.3 Procedure 140
  - 35.2.4 Calculations 140
  - 35.2.5 Precautions 140
- 35.3 Triglycerides Estimation 141
  - 35.3.1 Principle 141
  - 35.3.2 Reagents 141
  - 35.3.3 Procedure 142
  - 35.3.4 Precautions 142
- 35.4 HDL Estimation 142
  - 35.4.1 Principle 142
  - 35.4.2 Reagents 143
  - 35.4.3 Procedure 143
  - 35.4.4 Precautions 143
- 35.5 LDL Estimation 143
  - 35.5.1 Principle 143
  - 35.5.2 Reagents 144
  - 35.5.3 Procedure 144
- 35.6 Clinical Significance 144
  - 35.6.1 Cholesterol Levels 144
  - 35.6.2 LDL Values 145
  - 35.6.3 HDL Values 145
  - 35.6.4 Triglycerides 145

36 To Estimate Sodium and Potassium in Serum by Using Flame Photometer 147
- 36.1 Theory 147
- 36.2 Specimen Type, Collection, and Storage 147
- 36.3 Principle 148
- 36.4 Reagents 148
- 36.5 Procedure 149
- 36.6 Clinical Significance 149

37 To Perform Radioimmunoassay 151
- 37.1 Principle 151
- 37.2 Advantage 152
- 37.3 Applications 152

38 To Perform Enzyme-Linked Immunosorbent Assay 153
- 38.1 Principle 153
- 38.2 Sandwich ELISA 153
- 38.3 Competitive ELISA 154

	38.4	Direct ELISA	155
	38.5	Applications of ELISA	155
39	**Some Important Case Studies**		157
	39.1	Case Studies of Sugar Impairment	157
	39.2	Case Studies of Diabetic Ketoacidosis	158
	39.3	Case Studies of Calcium and Phosphate Impairments	159
	39.4	Case Studies of Protein Energy Malnutrition	160
	39.5	Case Studies of Gout/Uric Acid	160
	39.6	Case Studies of Liver Functions	162
	39.7	Case Studies of Kidney Functions	165
	39.8	Case Studies of Cardiac Functions	167

Suggested Readings ... 169

Index ... 171

About the Authors

K. D. Gill did her postgraduate in Biochemistry from Punjab Agricultural University, Ludhiana, Punjab, India, in 1969. She completed her doctorate degree from the same university in 1974 and then joined as lecturer in the Department of Chemistry, Punjabi University, Patiala, in 1976. Then she joined as postdoctoral fellow, at Physiological Chemistry Department, Faculty of Medicine, Gottingen, W. Germany, in 1977. In 1981, she was appointed as lecturer in Biochemistry at the Department of Biochemistry Postgraduate Institute of Medical Education and Research, Chandigarh, India, where she served in various capacities such as assistant professor (1986–1987), associate professor (1987–1991), additional professor (1991–2001), and professor (2001–2012). She became head of Biochemistry Department in 2012 and retired in that capacity. She has 32 years of teaching Biochemistry subject at postgraduate level. She has published more than 125 research papers in journals of international repute. She is fellow of the Society of Applied Biotechnology and has served as editorial board member and associate editor for many journals. She has four book chapters to her credit. She has also supervised clinical Biochemistry labs.

Vijay Kumar did his postgraduate in biochemistry from Maharshi Dayanand University, Haryana, India, in 2002. He then joined the Department of Biochemistry at Postgraduate Institute of Medical Education and Research, Chandigarh, to pursue his PhD and obtained his PhD degree in 2009. After getting his degree, he served there as demonstrator and worked in clinical biochemistry experimentation and also taught biochemistry to MD students. In 2010, he joined as assistant professor at the Department of Biochemistry, Maharshi Dayanand University, Haryana. He is actively engaged in teaching Biochemistry to postgraduate and PhD students.

Abbreviations

α	Alpha
&	And
ACP	Acid phosphatase
ADP	Adenosine diphosphate
ALP	Alkaline phosphatase
ALT	Alanine amino transferase
ANSA	1,2,4-Amino naphthol sulfonic acid
AST	Aspartate amino transferase
ATP	Adenosine triphosphate
$BaCl_2$	Barium chloride
β	Beta
BCG	Bromocresol green
BSA	Bovine serum albumin
CK	Creatine kinase
Conc	Concentrated
CSF	Cerebrospinal fluid
$CuSO_4$	Copper sulfate
DNPH	2,4-Dinitrophenylhydrazine
EDTA	Ethylenediaminetetraacetic acid
ELISA	Enzyme-linked immunosorbent assay
$FeCl_3$	Ferric chloride
γ	Gamma
g	Gram
GFR	Glomerular filtration rate
GOD	Glucose oxidase
GTT	Glucose tolerance test
H_2O_2	Hydrogen peroxide
H_2SO_4	Sulfuric acid
H_3PO_4	Phosphoric acid
HCl	Hydrochloric acid
HDL	High-density lipoprotein
HNO_3	Nitric acid
H_3PO_4	Phosphoric acid

h	Hour(s)
KCl	Potassium chloride
kDa	Kilodalton
L	Liter
LDH	Lactate dehydrogenase
LDL	Low-density lipoprotein
M	Molar
mg	Milligram (s)
min	Minute (s)
mM	Millimolar
mol	Mole(s)
m mol	Millimoles
Na_2CO_3	Sodium carbonate
Na_2HPO_4	Disodium hydrogen phosphate
NaCl	Sodium chloride
NAD	Nicotinamide adenine dinucleotide
NADP	Nicotinamide adenine dinucleotide phosphate
NaH_2PO_4	Sodium dihydrogen phosphate
$NaHCO_3$	Sodium bicarbonate
NaOH	Sodium hydroxide
nm	Nanometer
OD	Optical density
%	Percent
pI	Isoelectric point
POD	Peroxidase
RBCs	Red blood cells
RNA	Ribonucleic acid
RIA	Radioimmunoassay
Sec	Second(s)
TCA	Trichloroacetic acid
Tris	Tris(hydroxymethyl)aminomethane
UV	Ultraviolet
VLDL	Very-low-density lipoprotein
μM	Micromolar
μg	Microgram(s)
<	Less than
>	Greater than

Common Clinical Laboratory Hazards and Waste Disposal

1

It is very important to protect laboratory worker from hazards. The laboratory hazards fall into three main categories: chemical hazards, biological hazards, and physical hazards. The major chemical hazards in laboratories are the cleaning agents, anesthetic gases, disinfectants, drugs, and solvents. The persons working in laboratory are exposed to chemical hazards during their usage or due to improper storage. Biological hazards involve exposures to infectious samples, animal diseases transmissible to humans, and biological agents used during experimental procedures that include viral vectors, etc. The exposure to physical hazards is associated with research facilities. The laboratory personnel encounter the physical hazards due to accidental spill of corrosive reagent, broken glassware, etc. (Fig. 1.1).

The major precautions that can be followed while working in laboratory are:

1. Proper labelling of chemicals.
2. The worker should avoid mouth pipetting and should take care not to blow out pipettes containing potentially infectious material like serum.
3. The gloves, mask, protective eyewear, and gowns must be used while drawing blood from a patient.
4. Proper installation of fire extinguishers, grounding of electrical wires, etc.
5. The reuse of the syringes should be avoided, and needles should be disposed off in the containers without touching.
6. Completely incinerate infectious tissues.
7. The discarded tube or infected material should not be kept unattended or unlabelled.
8. All sharp objects and infectious samples should be disposed off properly.
9. Immediately use laboratory first aid in case of accidental exposure to any hazard (see Table 1.1).
10. The laboratory workers must be trained for proper handling and disposal of biohazardous materials including the patient specimens. (Note the hazard warning symbol on reagents.)

© Springer Nature Singapore Pte Ltd. 2018
V. Kumar, K. D. Gill, *Basic Concepts in Clinical Biochemistry: A Practical Guide*,
https://doi.org/10.1007/978-981-10-8186-6_1

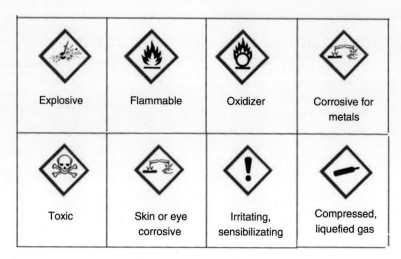

Fig. 1.1 Hazard warning symbols

Table 1.1 Laboratory first aids

S. no	Accident/injury	First aid
1.	Alkali splash on the skin	Wash with tap water for 15 min followed by 5% acetic acid solution
2.	Acid splash on the skin	Wash with tap water for 15 min followed by 5% sodium carbonate solution
3.	Phenol burn	Wash with plenty of tap water. Then use polyethylene glycol with water
4.	Splashes in the eyes	Wash with plenty of tap water and sterile saline. Then seek professional doctor's help
5.	Injury due to broken glass	Wash wound immediately with disinfectant
6.	Burn	Wash with plenty of tap water and cover with sterile dressing

1.1 Waste Disposal in Laboratory

Laboratory wastes may be hazardous in nature. It should be disposed off in appropriate plastic bin bags:

1. *Black waste bin bag* – Contaminated medicine, general waste, leftover food and peels of fruits, outdated medicine, any noninfectious material, etc. should be discarded in these waste bins.
2. *Red waste bin bag* – Plastic waste such as catheters, urinary catheter, suction catheter, Ryle's tube, injection syringes, tubing, IV bottles, used or discarded blood containers, microbiology culture, etc.

3. *Blue waste bin bag* – All broken glass bottles and articles, surgical blades, glass syringes, needles, and any sharp material.
4. *Yellow waste bin bag* – Empty vials, ampoules, gloves, infectious waste, human anatomical wastes, organs, body parts, dressing, bandages, gauze, items contaminated with blood/body fluids, microbiological and biotechnical wastes, etc.

Blood Collection and Preservation

2

2.1 Blood Collection

Blood is a body fluid containing plasma, red blood cells (RBCs), white blood cells, and platelets. Blood is specialized for performing various functions such as transport of nutrients and oxygen to various body organs, transportation of antibodies, transport of waste products to kidneys, and regulation of body temperature. Normally, blood pH is maintained in narrow range of 7.35–7.45. Blood for biochemical investigations may be drawn from arteries, veins, or capillaries. Venous blood is commonly used for majority of biochemical investigations. It can be drawn from any prominent vein. Arterial blood is mostly used for blood gas analysis. Radial, brachial, or femoral arteries are the most common site for arterial blood. Capillary blood is collected by puncture in infants or when very little blood is required. Blood is collected in various collection tubes called vacutainers which are sterile glass tubes with a colored rubber stopper. Blood contains various chemical constituents such as glucose, proteins, lipids, globulin, fibrinogen, urea, amino acids, uric acid, creatinine, hormones, vitamins, electrolytes, etc. (Fig. 2.1).

Hemolysis of Blood Hemolysis is the release of hemoglobin from red blood cells that give pink to red color to plasma/serum. Hemolysis during sampling, transportation, and storage should be avoided since it causes changes in measurement of a number of analytes.

Collection of Specimen Many factors need to be considered when collecting lab specimen. In some cases, preparation of the patient prior to the test may be required. The sample volume to be collected depends upon the number and type of tests being performed. Generally 3–5 ml blood is required for many investigations. If the tests are run on automated instruments, then less volume of blood may be sufficient.

Fig. 2.1 Vacutainers used for blood collection and storage

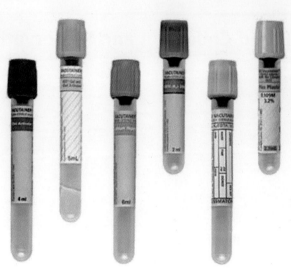

Plasma Separating Tubes These tubes contain ethylenediaminetetraacetic acid (EDTA) which is a strong anticoagulant. The tubes should be inverted several times after sample collection.

Procedure for Plasma Preparation
Draw blood from patient and pour it in vacutainer with an appropriate anticoagulant. Mix blood with anticoagulant properly and allow the tubes to stand for 10 min. Then, the sample is centrifuged for speed separation and packing of cells. The supernatant is the plasma.

Serum Serum composition is the same as that of plasma except that serum lacks fibrinogen.

- For many laboratory biochemical tests, plasma and serum both can be used interchangeably.

Procedure for Serum Preparation
1. Draw blood from patient. Select vacutainer without anticoagulant.
2. Allow the vacutainer to stand for 20–30 min so that clot is formed.
3. Centrifuge the sample at 3000 rpm which affects a greater packing of cells. Various cells along with clot will settle in the form of pellet at the bottom of the tube.
4. The supernatant is the serum.

Anticoagulants

Whole blood or plasma sample investigations need use of anticoagulants while collecting sample. Some common anticoagulants are:

(a) *Ethylenediaminetetraacetic acid*: This anticoagulant is used at a concentration of 2 mg/dl of blood volume. It removes calcium ions by chelation and block coagulation. It is used mainly for hematological studies.

(b) *Heparin*: Heparin inhibits conversion of prothrombin to thrombin. Heparin is present naturally in blood and hence acts as ideal anticoagulant. It increases the activity of antithrombin. For every ml of blood sample, 0.2 ml of heparin may be used.

(c) *Sodium fluoride*: This anticoagulant is considered when glucose estimations are carried out in blood samples. Sodium fluoride inhibits glycolysis by inhibiting activity of enolase enzyme and hence preserves blood glucose levels. It is generally combined with potassium oxalate because of its poor anticoagulant action.

(d) *Sodium or potassium oxalate*: Sodium, potassium, and even lithium oxalates precipitate calcium ions and inhibit blood coagulation. Potassium oxalate is more water soluble and is used at concentration of 5–10 mg/5 ml of blood.

Quality Control in Laboratory

The ultimate goal of the clinical biochemistry laboratory is to analyze the substances in body fluids or tissues both qualitatively and quantitatively for diagnosis and treatment of disease. The presentation of incorrect laboratory results may lead to wrong diagnosis and treatment leading to fatal results. Hence, it is very important to generate the reliable data that depends on strict quality control management.

Quality control is the procedures of corrective responses employed for the detection and measurement of the sources of variation or errors. In simple words, we can present that quality control is a representation of precision and accuracy under varying experimental conditions. The various criteria included for reliable analytical methods are:

Accuracy It is the degree of agreement between large numbers of measurements on a sample with the actual quantity of a substance present in the sample. Accuracy depends upon the methodology used for sample measurement.

Precision Precision refers to the reproducibility between repeated determinations of an analyte. The precision depends on accuracy of the methods used for sample analysis.

Specificity It is the ability of an analytical method to discriminate between similar substances being analyzed.

Sensitivity Sensitivity is the capacity of an analytical method to measure the minimum quantities of analytes under consideration.

To maintain the criteria for reliability in analysis, calibration is done regularly. Calibration is performed by using the various standards. A standard is the solution with known amount of analytes and with which the sample can be compared to derive the result.

While performing analytical measurements in laboratory, various types of errors may be encountered.

3.1 Types of Laboratory Errors

There are three major errors that may occur in a laboratory:

1. *Random errors*: Random errors are the errors that arise due to statistical fluctuations in the observations and lead to inconsistent measurement value of a constant attribute. A random error is associated with the fact that when a measurement is repeated, it will generally provide a measured value that is different from the previous value. Random errors are caused by uncontrollable variables, which cannot be defined or eliminated. These are errors that may arise due to bubbles in reagents or reagent lines, instrument instability, temperature variations, and operator variability, such as variation in pipetting.
2. *Systemic errors*: Systematic errors cause inaccurate results that are consistently low or high. This error is reproducible and predictable and can be easily identified and corrected. These errors are caused by insufficient control on analytical variables, e.g., impure calibration material, change in reagent lot, change in calibration, assigning the wrong calibrator values, improperly prepared or deteriorating reagents, etc. Majorly, systemic errors arise due to three factors:
    (a) *Instrument errors*: Instrument errors are errors associated with instrument functioning. These arise due to power fluctuations, defect in any parts of the instrument, temperature variation, or when the instrument is not calibrated. The instrumental errors can be removed by proper calibration or maintenance of instrument.
    (b) *Method errors*: These are errors that arise due to the use of non-ideal physical or chemical methods. For example, the speed of reaction, problem associated with sampling, and interference from side reactions can lead to such errors. The development and use of proper method can help to correct these errors.
    (c) *Personal errors*: These are caused by an observer's personal habits or mental judgment, wrong judgment of dimensional values, color acuity problems, etc. It may be accidental or systematic. Proper training and experience can help to eliminate the personal errors effectively.
3. *Gross errors or total analytical error*: Such errors arise due to equipment failure or observer's carelessness.

The laboratory errors may also be grouped into pre-analytical, analytical, or post-analytical errors according to time of occurrence. *Pre-analytical errors* arise before the analysis of sample takes place. Common examples of pre-analytical errors are mismatch of sample and laboratory data, error in presentation of analyzed results, and delaying in report generation. The *analytical errors* occur during analytical methods and include errors related to expired or spoiled reagents, use of controls or calibrators that have expired, sampling errors, and changes in analyzer's measuring unit. *Post-analytical errors* arise during transmission of data from analyzers, result validation, and dispatching/communicating results to physicians or patients. Common post-analytical errors are loss of the results, inappropriate specimen or anticoagulant, error in storage of sample, or mistakes in patients' identification.

3.2 Methods to Minimize the Laboratory Errors

The routinely used methods being performed in a laboratory should be monitored continuously for any change in precision or accuracy. The quality control techniques must be used regularly to detect such changes and to follow corrective measures. It is always desirable that both random and systematic errors must be detected at the earliest possible stage and preventive measures should be applied to minimize them.

The procedure adopted for the detection of errors consists of specific quality control methods. Such methods can be divided in two categories:

Internal Quality Control The internal quality control involves internal processes followed for monitoring of experimental protocols and checking of the resultant data systematically. Internal quality control mechanism is necessary to reach the conclusion that processed data is reliable enough to be released. In other words, it correlates the day-to-day quality control methods with the equipment and methods used in the laboratory. The process predicts about the precision and accuracy of the method being followed. This process also involves construction of Levey-Jennings chart and quantification of unavoidable laboratory data (Fig. 3.1).

The internal quality control considers the following points:

1. The correlation of clinical test being used with the diagnosis of disease.
2. The analysis of the same sample should be performed two times, and identical results must be obtained if no error exists.
3. The reproducibility of samples must be checked frequently.

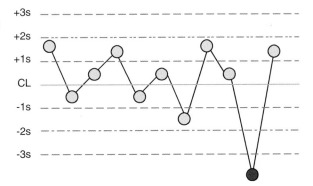

Fig. 3.1 Levey-Jennings chart with a random error and a systematic error

4. If a test is conducted on patient the second time, then the results obtained from such test may be compared with the previous results. The values may be increased with progression of disease or vice versa.

External Quality Control The external quality control involves monitoring the resultant outcome of clinical tests by using controls or reference samples provided from other sources. Such analysis is done without reference values for analyte under the conditions of that lab. The quality control methods are performed at regular intervals by the laboratory personnel by using the controls/standards of an external referral laboratory, without having any idea of resultant value. It is helpful to check primarily the accuracy of analytical methods of any laboratory. If the results for any methods which are followed in a laboratory show deviation from other established methods, then the methods should be replaced immediately by another after revaluation of other parameters such as the use of calibration standards, reagents, pipettes, and instruments used for measurements.

Automation in Clinical Laboratory

4

Automation is the use of various control systems for operating equipments and other applications with minimum human intervention. The use of automation in clinical laboratory enables to perform many tests by analytical instruments with minute use of an analyst. The automated instruments have advantages that laboratories can process more workload with minimum involvement of manpower. Also, automation minimizes the chances of variability of results and errors that generally can occur during manual analysis. Although reproducibility has improved due to the use of automation in recent years, this may not necessarily be associated with more accuracy of test results. In fact the accuracy is influenced by the analytical methods used. The automated analyzers are the mechanized versions of basic manual laboratory techniques, and recently, many analytical methods with modification of existing protocols have been developed which are fast and easy to operate. The use of analyzers with integrated computer hardware and software has made the job very easy for clinical laboratories as it provides automatic process control and data processing. In simple words, the credit for significant improvement in the quality of laboratory tests in progressive years may be attributed to well-designed automated instrumentation, improved analytical methods, and effective quality assurance programs. The use of automated analyzer has many advantages including reduction of workload, less time consumption per sample analysis, more number of tests done in less time, use of minute amount of sample, decreased chances of human errors, and high accuracy and reproducibility.

4.1 Types of Autoanalyzers

There are various variants of autoanalyzers.

Semi-autoanalyzer In semi-autoanalyzer, pipetting of samples, reagent mixing and incubation, etc. are done manually. The ready sample is kept in autoanalyzer to read

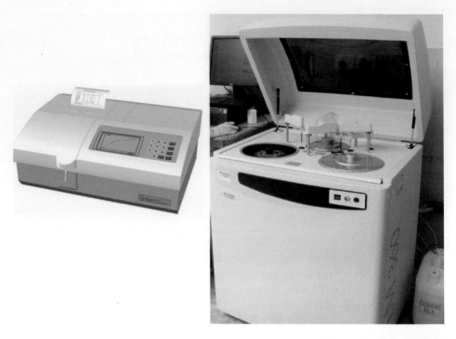

Fig. 4.1 Semi-autoanalyzers

sample. These are usually meant to measure one analyte at a time, but many samples may be measured rapidly.

Batch Analyzer In batch analyzer, one batch of a specific test is performed at a time automatically by the analyzer. In these instruments, reagent mixture is prepared and fed automatically, and one reagent is stored in the machine for running a test.

Random Access Autoanalyzers In such analyzers, more than one reagent is stored at a time. The samples are placed in the machine, and it can perform any number of specific tests on each sample (Figs. 4.1 and 4.2).

The steps followed for analysis of sample on the automated systems:

1. *Identification of sample*: The tubes containing blood or other fluid samples for analysis are labelled at collection center. On reaching the lab, the description of sample is recorded in computers, and then they are processed.
2. *Bar coding*: The technology of bar coding is used by many clinical laboratories for the identification of sample. Several analytical systems are equipped with bar coding facility. A bar reader present in such systems reads and records the sample information which is transferred to the system and processed by software.
3. *Preparation and transport of sample*: The sample is prepared before the analysis is performed. Various steps like blood clotting, centrifugation of sample, and

4.1 Types of Autoanalyzers

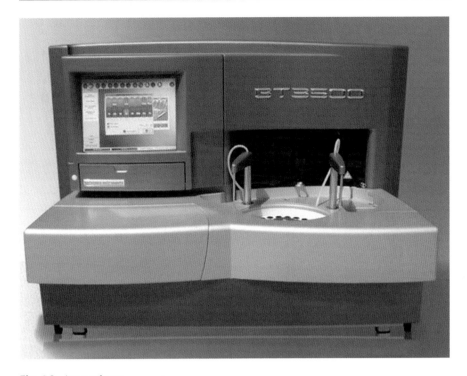

Fig. 4.2 Autoanalyzer

separation of serum are performed to enable sample preparation. The use of whole blood for sample analysis and automation of specimen can be done to process the sample faster. Now, the sample tubes are transferred to the analyzer room and loaded on the analyzer.

4. *Processing of sample*: The sample is processed before the measurement is done. The processing includes the removal of the interfering substances from sample for accurate testing on the analyte.
5. *Reagent preparation and loading*: All reagents used for analysis should be stored at 4°C until usage. The reagents are mixed if required and loaded on analyzer.
6. *Sample measurement*: During analysis on autoanalyzer, the samples react with reagents and undergo chemical reactions under optimum conditions. The measurements are processed automatically, and output signals are sent in the form of results.

Photometry: Colorimeter and Spectrophotometer

Photometry deals with the study of phenomenon of light absorption by molecules in solution. The specificity of a compound to absorb light at a particular wavelength is useful in quantitative measurements. When a light beam of particular wavelength is passed through a solution, some amount of light is absorbed by the solution, and consequently the intensity of light that comes out of solution is diminished. The phenomenon of absorption of light by a solution follows Beer-Lambert's law. Beer's law states that amount of transmitted light decreases exponentially with increase in the concentration of absorbing medium. In other words, absorbance is directly related to the concentration of the absorbing medium. Lambert's law states that amount of transmitted light decreases exponentially with increase in thickness of absorbing medium. The two laws can be combined, hence

$$I = I_0^{\varepsilon c l}$$

where

I = Intensity of transmitted light
I_0 = Intensity of incident light
ε = Molar absorption coefficient
c = Concentration of absorbing substance (g/dl)
l = Thickness of absorbing medium or solution

When thickness of absorbing medium is kept constant, then intensity of incident light depends only upon the concentration of absorbing medium, or we can say that Beer's law is operative. The ratio of transmitted light to incident light is called transmittance, whereas absorbance or optical density can be calculated as

$$\text{Absorbance} = 2 - \log_{10} T$$
$$= 2 - \log\% \ T$$

5.1 Colorimeter and Spectrophotometer

Colorimeter and spectrophotometer work on principle of light absorption which is integral part of a large number of instruments including semi-autoanalyzers and autoanalyzer. Colorimeter is an instrument used to measure the concentration of colored substances. It works in visible range of 400–800 nm of electromagnetic spectrum of light. The working of colorimeter is based on Beer-Lambert law and is based on the fact that any metabolite or chemical substance that undergoes a reaction to form a colored product in the cuvette will absorb some amount of light. The amount of the light absorbance will be proportional to the concentration of the metabolite in the sample. The functioning of colorimeter is simple. The light source which is generally a filament lamp emits light. The filter blocks all wavelengths of light source except the desired wavelength at which sample is read. This light with selective wavelength then passes through a sample holder which holds a special glass cuvette with a fixed thickness. Some portion of incident light is absorbed by the sample in cuvette, and the light that is not absorbed by the sample (also called transmitted light) falls on photoelectric detectors. These photoelectric detector cells measure the intensity of incident light falling on it and calculate the amount of absorbed light or absorbance (Fig. 5.1).

Spectrophotometer It also measures absorbance of light intensity by a chemical substance when incident light passes through sample solution. The basic principle of colorimeter and spectrophotometer is the same except that spectrophotometer also covers ultraviolet (UV) region range (200–400 nm) in addition to visible range of electromagnetic spectrum. Unlike recording the absorbance readings at a set wavelength as used in colorimeter, the absorbance can be recorded over a range of

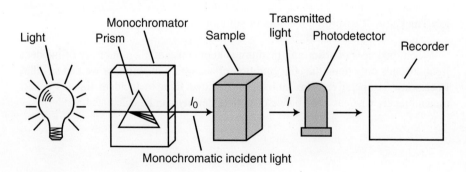

Fig. 5.1 Basic components of colorimeter

Fig. 5.2 Spectrophotometer

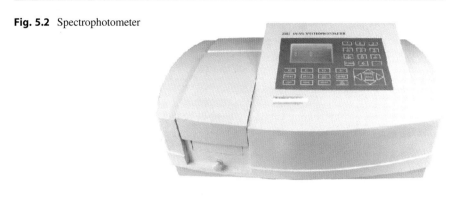

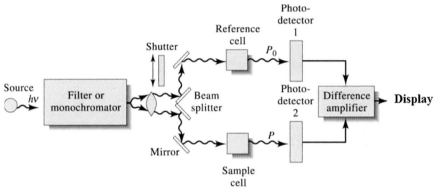

Fig. 5.3 Working principle of double beam spectrophotometer

wavelengths in spectrophotometer even at every 5 or 10 nm ranges. It gives a data spectrum instead of giving a specific absorbance reading.

The spectrophotometers are of two types – single beam and double beam. In single beam spectrophotometers, there is single beam from light source. The reference sample also called blank sample is first kept in sample holder, and wavelength absorption is measured. Then it is removed, and sample under consideration is placed in sample holder, and absorption of light is measured again. The absorbance value of reference sample is subtracted from the absorbance value of samples under consideration. This process removes the additive effects from the solvent and the cell. In double beam spectrophotometer, the light from source is split in two beams. One beam illuminates the reference or blank sample, and other beam passes through sample under consideration. These beams may be recombining before they reach monochromator. In some cases, two monochromator may be used. The display shows measurement of reading as difference of reference and test sample. In these instruments, 2–6 samples may be read in single run (Figs. 5.2 and 5.3).

The sample cuvette used in spectrophotometer is made of quartz instead of glass cuvette as in colorimeter. Although glass cuvette can be used for wavelength measurement in visible range, but only quartz cuvette can be used for wavelength measurement in UV region since glass absorbs UV light. The light source used is deuterium lamps for the UV measurement, whereas tungsten or halogen lamp is used for measurement in visible range of electromagnetic spectrum. These days, xenon arc lamps are also used which produces light of both UV and visible spectrum.

Preparation of General Laboratory Solutions and Buffers

Almost all the experimental techniques in clinical biochemistry require the use of solutions. A solution is a homogenous mixture of two or more nonreactive substances. It is made up of two components. The dissolved substance is called solute, and the medium in which solute is dissolved is called solvent. The solutions are prepared in terms of concentrations. The concentration of a solution can be defined as the amount of solute present in a given quantity of solution. Following are the common ways to express the concentration of a solution:

6.1 Molar Solutions

One molar solution (1 M) contains 1 g molecular weight of solute per liter of solution. For example, 1 M solution of NaOH contains 40 g of NaOH per liter of solution.

6.1.1 Molarity (M)

Molarity of a solution, also termed as molar concentration, is the number of moles of solute dissolved in 1 litre of solution. The molarity of a solution can be calculated when the number of grams of the substance present or dissolved in the solution, molecular weight of the dissolved substance, and the volume of the solution are known.

$$\text{Molarity (M)} = \frac{\text{Number of moles of solute}}{\text{Litre of solution}} = \frac{\text{Weight in gram} \times 1000}{\text{Molecular weight} \times \text{volume (ml)}}$$

For example, the molarity of a 1 litre solution containing 4 g of NaOH will be 0.1 M, i.e.,

$$\text{Molarity} = \frac{4 \times 1000}{40 \times 1000} = 0.1 \text{M}$$

If we know the molarity of solution to be prepared, then the amount (weight) of salt that needs to be dissolved for preparation of per liter solution can be calculated.
For example, to prepare 1 litre solution of 0.2 M NaOH,
Molecular weight of NaOH = 40, so

$$0.2 = \frac{\text{Weight in gram} \times 1000}{40 \times 1000}$$

Weight in gram = $0.2 \times 40 = 8$ g (i.e., 8 g of NaOH is dissolved in 1 litre final volume of distilled water).

Alternatively, if we directly multiply required molarity with molecular weight, it will give weight of salt in gram per liter of solution.
For example, to calculate amount of salt to prepare 0.2 M NaOH,

$$0.2 \times 40 = 8 \text{ g/L}$$

Note If required molarity is in mM, then salt amount will be in mg/liter, and if molarity is in μM, then amount will be in μg/liter.

However, the molarity of conc. solution can be expressed as

$$M = \frac{\%\text{purity} \times \text{specific gravity} \times 1000}{\text{Molecular weight} \times 100}$$

6.2 Normal Solutions

One normal solution is the solution which contains 1 g equivalent weight of solute per liter of solution.

6.2 Normal Solutions

6.2.1 Normality (N)

Normality can be defined as the number of equivalents of a solute dissolved in 1 litre of solution. An equivalent is the molecular mass of the acid or base expressed in grams divided by the number of moles of hydronium or hydroxyl ions produced by this amount of acid or base.

$$N = \frac{\text{Molecular mass} \times 1000}{\text{Equivalent weight} \times \text{volume (ml)}}$$

The equivalent weight of an acid can be calculated as

$$\text{Equivalent weight} = \text{molecular weight of an acid/basicity}$$

The basicity of an acid is defined as the number of replaceable hydrogen atoms present in one molecule of acid. That is, for hydrochloric acid (HCl), molecular weight is 36.5, and basicity is 1, so equivalent weight is 36.5; however, in case of sulfuric acid (H_2SO_4), molecular weight is 98, and basicity is 2, so equivalent weight will be 98/2 = 49.

The equivalent weight of a base can be calculated as

$$\text{Equivalent weight} = \text{molecular weight of an base/acidity}$$

The acidity of a base is defined as the number of hydroxyl ions which a molecule of base can furnish in aqueous solution. For example, acidity for NaOH is 1, and for Na_2CO_3, it is 2.

The normality of given conc./stock solution can be calculated as

$$N = \frac{\%\text{purity} \times \text{specific gravity} \times 1000}{\text{Equivalent weight} \times 100}$$

If we have to prepare a solution of specific normality from a given stock solution of acid/alkali with known normality, then such solutions can be prepared by applying the relationship given below:

$$N_1 \times V_1 = N_2 \times V_2$$

where

N_1 = normality of stock solution
V_1 = volume of stock solution
N_2 = normality of solution to be prepared
V_2 = volume of solution to be prepared

6.3 Percent (%) Solutions

The % solutions can be expressed as below:

(i) Weight/volume (%, w/v): It is the weight of a solute in grams dissolved in 100 ml of solution, i.e., 4 g pellet of NaOH is dissolved in 100 ml final volume of distilled water.
(ii) Volume/volume (%, v/v): The volume of a solute in ml dissolved in 100 ml of solution, i.e., 20% ethanol (v/v) means that 20 ml of ethanol is mixed with 80 ml of distilled water to make final volume 100 ml.

6.4 Buffer Solutions

A buffer is defined as the solution that resists changes in the pH upon addition of small amount of acid or base. A buffer is an aqueous solution of a weak acid and its conjugate base or weak base and its conjugate salt. The role of buffers in maintaining the pH of a solution can be explained with the aid of the Henderson-Hasselbalch equation.

$$pH = pKa + \log \frac{[\text{conjugate base}]}{[\text{conjugate acide}]}$$

The conjugate acid and conjugate base act together to resist large changes in pH by partially reacting with H^+ or OH^- ions added to the buffer solution. Let us consider the example of acetate buffer that contains acetic acid and sodium acetate. Acetic acid ionizes weekly, and sodium acetate ionizes to a large extent in solution. When H^+ ions (i.e., HCl) are added to this buffer solution, the acetate ions present in buffer bind with H^+ to form acetic acid, and the H^+ ions are taken out of circulation. The acetic acid is ionized very weakly, and hence, pH change due to addition of HCl is resisted by buffer. Similarly, if OH ions (i.e., NaOH) are added to buffer solution, the H^+ ions present in buffer combined with OH^- ions, and OH^- ions are taken out of circulation, and pH change is again resisted by buffer. The amount of change in pH of buffer depends on the strength of the buffer and the ratio of conjugate base/conjugate acid. The efficiency of a buffer to resist changes in pH on addition of acid or base is called as the "buffer capacity." Buffer capacity can be defined as the number of moles of H^+ or OH^- that must be added to 1 litre of the buffer in order to change the pH by one unit. The buffering capacity of the buffer solution is maximum when pH = pKa, i.e., when the concentrations of conjugate acid and conjugate base are equal. In general, buffers should not be used at a pH greater or lower than 1 pH unit from their pKa. The most commonly used buffers in the laboratory are acetate buffer, phosphate buffer, citrate buffer, and Tris buffer.

6.5 Physiological Buffers in the Human Body

In the first instance, pH in the human body is maintained by physiological buffers. Buffers may be intracellular and extracellular. Different buffer systems work in correlation with one another. It means that changes in one buffer system lead to changes in another. Various metabolic processes in body produce substantial amounts of acids and bases. The production of these acids and bases may disturb the blood pH. The normal blood pH is 7.4, and it is regulated very precisely. Various conditions such as uncontrolled diabetes mellitus, nephritis, vomiting, diarrhea, etc. cause change in blood pH. The physiological buffers help effectively to regulate blood pH and maintain acid-base balance of body which if disturbed can disturb metabolic processes.

The main physiological buffers in our body are the following:

1. **Carbonate-bicarbonate buffer**: It is the most important extracellular buffer and has the largest buffering capacity. It is present in high concentration in plasma and acts in cooperation with other buffers. It also plays an important role in the red blood cells.
2. **Hemoglobin buffer**: It is the main intracellular buffer of the blood and regulates blood pH by the removal of hydrogen ions from blood cells.
3. **Protein buffer**: It is an extracellular buffer together with bicarbonate buffer, represented by plasma proteins. The histidine content of proteins plays an important role in its buffering ability due to its pKa very near to blood pH.
4. **Phosphate buffer**: This buffer takes part in hydrogen ion excretion in renal tubules and is not of great importance in blood due to its less concentration although it has pKa close to blood pH.

6.6 Preparation of Common Laboratory Buffers

6.6.1 0.2 M Acetate Buffer (pKa 4.86)

Acetate buffer is prepared by the mixing of acetic acid and sodium acetate.

Prepare 0.2 M acetic acid and 0.2 M sodium acetate solutions separately. Add two solutions in different proportions to obtain various pH solutions.

S. no	Volume of 0.2 M acetic acid taken	Amount of 0.2 M acetic acid in used volume (μ mole)	Volume of 0.2 M sodium acetate taken (ml)	Amount of 0.2 M sodium acetate in used volume (μ mole)	Theoretical pH
1.	46	9200[a]	4	800	3.8[a]
2.	42	8400	8	1600	4.1
3.	38	7600	12	2400	4.3
4.	34	6800	16	3200	4.5
5.	30	6000	20	4000	4.6

[a]Since 1000 ml of 0.2 M acetic acid contain = 0.2 moles,
so 46 ml of 0.2 M acetic acid will contain = 0.2 × 46/1000 = 9200 µmoles

Calculate pH using Henderson-Hasselbalch equation

$$pH = pKa + \log \frac{[\text{conjugate base}]}{[\text{conjugate acid}]}$$
$$pH = 4.86 + \log 800/9200$$
$$= 4.86 - 1.06 = 3.8$$

6.6.2 0.2 M Sodium Phosphate Buffer (pKa 6.86)

Sodium phosphate buffer is prepared by mixing sodium dihydrogen phosphate (NaH_2PO_4) and disodium hydrogen phosphate (Na_2HPO_4). Prepare 0.2 M solutions of monobasic and dibasic salts separately and mix dibasic solution in monobasic solution as given in table.

S. no	Volume of 0.2 M NaH_2PO_4 taken	Amount of 0.2 M acetic acid in used volume (µ mole)	Volume of 0.2 M Na_2HPO_4 taken	Amount of 0.2 M Na_2HPO_4 in used volume (µ mole)	Theoretical pH
1.	46	9200[a]	4	800	5.7[a]
2.	42	8400	8	1600	6.1
3.	38	7600	12	2400	6.3
4.	34	6800	16	3200	6.5
5.	30	6000	20	4000	6.6

[a]Since 1000 ml of 0.2 M NaH_2PO_4 contain = 0.2 moles,
so 46 ml of 0.2 M NaH_2PO_4 will contain = 0.2 × 46/1000 = 9200 µmoles

Calculate pH using Henderson-Hasselbalch equation

$$pH = pKa + \log \frac{[\text{conjugate base}]}{[\text{conjugate acid}]}$$
$$pH = 4.86 + \log 800/9200$$
$$= 6.86 - 1.06 = 5.7$$

6.6.3 0.2 M Tris-HCl Buffer (pKa 8.1)

Tris buffer is prepared by mixing appropriate volumes of Tris and HCl as given in table.

6.6 Preparation of Common Laboratory Buffers

S. no	Volume of 0.2 M Tris taken	Amount of 0.2 M Tris in used volume (μ mole)	Volume of 0.2 M HCl taken	Amount of 0.2 M HCl in used volume (μ mole)	Volume of distilled water added (ml)	Theoretical pH
1.	50	10,000[a]	5	1000	45	9.1[a]
2.	50	10,000	10	2000	40	8.8
3.	50	10,000	15	3000	35	8.6
4.	50	10,000	20	4000	30	8.5
5.	50	10,000	25	5000	25	8.4

[a]Theoretical pH can be calculated as explained for acetate and phosphate buffer.

Examination of Urine for Normal Constituents

7

Urine is the excretory product of the body produced by the process of filtration, reabsorption, and tubular secretion. Urine can be collected and examined easily, and presence of certain substances in the urine may indicate the metabolic state of the body. The routine and microscopic examinations of urine are helpful in the diagnosis of several pathological conditions. Urine sample is collected in clean vials. A random sample is usually taken for routine clinical examination; however, first-morning urine is preferred for urinalysis and microscopic analysis, since it is generally more conc. Timed collection specimens may be required for quantitative measurement of certain analytes. Fresh or preserved urine samples are used for chemical and culture-based microbiological testing. The specimens that are unpreserved for more than 2 h or refrigerated for a long time may not be suitable for analysis due to potential bacterial overgrowth and invalidation of bacterial colony counts or errors in chemical urinalysis.

7.1 Preservatives Used for Urine Collection

The commonly used preservatives for urine sample collection in clinical biochemistry laboratory are:

1. 6 N HCl in proportion of 10 ml/24 h urine
2. Few drops of formalin in 30 ml of urine
3. 50% acetic acid
4. 6 N HNO_3
5. Boric acid, toluene, Na_2CO_3

7.2 Physical Examination of Urine

7.2.1 Color and Odor

Normal urine is colorless to straw colored due to presence of urochrome. Slight change in color occurs in fever, dehydration jaundice, or vitamin B-complex therapy which adds riboflavin (deep yellow color). Red to brown color is observed in hematuria, hemoglobinuria, myoglobinuria, and porphyria. Urine turns brown to black in alkaptonuria and methemoglobinuria. Normally odor of urine is faintly aromatic. On decomposition, a very unpleasant ammoniacal odor evolves. Food beverages and drugs may impart a specific odor to urine.

7.2.2 Appearance

Normally freshly voided urine is clear and transparent, but it may become turbid if exposed for a long time due to the bacterial action on urea present in urine to convert it into ammonium carbonate. Phosphate excretion in alkaline urine also makes urine turbid. The presence of white cells, red cells, or epithelial cells makes urine cloudy. Fat globules give urine milky appearance.

7.2.3 Specific Gravity

Specific gravity of urine is measured by urinometer. Specific gravity of normal urine is between 1.002 and 1.026 and depends upon state of hydration, diet, fluid intake, drugs, etc. Severe dehydration, diabetes mellitus, adrenal insufficiency, diabetes insipidus, and chronic nephritis increase specific gravity (Fig. 7.1).

7.2.4 Volume

Normal healthy individual excretes about 800–2000 ml of urine/day. Daily urinary excretion depends upon intake of fluid volume, loss of fluid, solute load, climatic condition, fever, or intake of drugs. The term polyuria is used if urinary output is more than 3000 ml/day. It occurs in conditions like diabetes insipidus, diabetes mellitus, or recovery from acute renal failure. In oliguria, urine output is less than 500 ml/day. Oliguria may be due to fewer intakes of water or due to dehydration or may indicate early renal dysfunction symptoms. In anuria, a complete cessation of urine output (<100 ml/day) is observed. Anuria should be taken care immediately; otherwise complete renal failure may take place.

Fig. 7.1 Urinometer

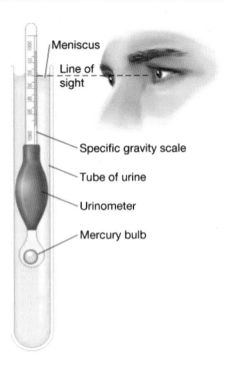

7.2.5 pH

Normal range of urinary pH is 4.5–7.5. Various factors like heavy meals, heavy exercise, metabolic acidosis, or chronic respiratory acidosis influence urine pH greatly. Normally freshly voided urine is acidic.

7.3 Chemical Examination of Urine

Normal urine has various organic and inorganic constituents. The normal inorganic constituents are chloride, phosphates, sulphates, calcium, etc., while organic constituents are urea, uric acid, and creatinine.

7.4 Tests for Inorganic Constituents of Urine

Chloride Add few drops of 3% silver nitrate solution to about 3 ml of urine sample. White precipitates appear due to the presence of chloride, carbonate, or phosphates. Addition of 3–5 drops of conc. HNO_3 leaves white precipitate of silver chloride only and dissolves carbonates and phosphates.

Phosphates To 3 ml of urine, add 1 ml of conc. HNO_3 and 5 ml of ammonium molybdate and warm gently. A yellow precipitate of phosphomolybdic acid indicates the presence of phosphate.

Sulphates To 3 ml of urine, add 3–5 drops of 2% barium chloride ($BaCl_2$) solution followed by 5 drops of conc. HCl. A white precipitate of $BaSO_4$ indicates inorganic sulphate.

Ammonia Boil urine with equal volume of 10% NaOH. Smell of ammonia gas indicates the presence of ammonium ion in the urine sample.

Calcium Add few drops of glacial acetic acid to about 5 ml of urine followed by addition of 1 ml of 4% ammonium oxalate solution. Calcium is slowly precipitated as oxalate as enveloped shaped crystals that can be observed under microscope.

7.5   Tests for Organic Constituents

Urea Since urea is the end product of protein metabolism, urea excretion in urine is an index of protein intake in diet. Urea in urine is increased in exaggerated protein metabolism, fevers, and increased adrenocortical activity. It is decreased in the last stages of severe hepatic diseases and in acidosis.

7.6   Detection of Urea

Hypobromite Test To about 3 ml of urine, add two drops of freshly prepared alkaline sodium hypobromite solution. A marked effervescence due to evolution of nitrogen indicates urea.

Biuret Formation Test Evaporate 3 ml of urine to dryness in a test tube to evolve ammonia. Cool and dilute the residue in 10% NaOH. Then add 1–2 drops of 1% $CuSO_4$. A red rose color formation indicates formation of biuret from urea on heating.

Creatinine Detection by Jaffe's Picric Acid Test Add 3 drops of saturated picric acid solution to about 3 ml of urine. Make it alkaline with few drops of NaOH solution. The red color formation confirms formation of creatinine picrate.

Uric Acid Detection To about 3 ml urine sample saturated with Na_2CO_3, add 2 drops of Folin's phosphotungstic acid reagent. Uric acid reduces phosphotungstate in alkaline solution and forms blue color.

To Perform Qualitative Tests for Urinary Proteins

8

8.1 Theory

Proteins are organic compounds containing nitrogen in addition to carbon, hydrogen, and oxygen. Sixteen percent of total protein weight is contributed by nitrogen. Proteins are responsible for distribution of body fluids and ions on both sides of the membrane. The plasma proteins act as buffers and regulate change in plasma pH. The normal protein excretion in urine is about 30–150 mg/day. The proteins in urine are normally derived from plasma filtrate and the lower lining of urinary tract due to tissue damage. The serum globulins, albumin, and proteins secreted by the nephron forms normal urinary proteins. About one third of urinary protein is albumin and remaining includes many small globulins. Plasma proteins of molecular weight less than 50 kDa easily pass through glomerular membrane of nephron and are usually reabsorbed from kidney tubular cells. Albumin with molecular weight of 66 kDa is filtered only in very small amounts. Retinol-binding proteins, β_2-microglobulins, immunoglobulin light chain, and lysozyme are also excreted in small amount. In summary, a healthy man excrete trace amount of proteins in urine. The large excretion of proteins in urine is called proteinuria. Proteinuria is an important indicator of kidney disease and the risk of disease progression. The proteinuria may also be caused by overflow of abnormal proteins in disease conditions such as multiple myeloma. The prevalence of proteinuria increases with kidney disease progression. The analysis of protein-to-creatinine ratio on a spot urine sample is used commonly to assess the 24 h urine protein excretion, and an albumin-to-creatinine ratio is used to approximate 24 h urine albumin excretion. Albumin-to-creatinine ratio is more sensitive than protein-to-creatinine ratio in detecting low levels of proteinuria. The normal urine albumin-to-creatinine ratio is less than 3.0 mg/mmol (ratio greater than 30 indicate severe increase of albumin excretion in urine), while normal protein-to-creatinine ratio is 15 mg/mmol.

8.2 Tests for Urinary Proteins

8.2.1 Dipstick Test

This is the most common test for proteinuria. Dipsticks are plastic strips impregnated with tetrabromophenol blue buffered to pH 3.0 with citrate. The test strips remain yellow in the absence of protein but changes to blue through various intermediate shades of green in presence of increasing protein concentration. This is due to divalent anionic form of indicator dye combining with proteins causing further dissociation of yellow monovalent anion into blue divalent anion. The blue-green color produced is directly proportional to the concentration of proteins in the specimen.

8.2.1.1 Method
Dip the strip in the urine for a second and remove the excess urine by tapping the edge against the container. Compare the color with the test chart within 30–60 s. The dipstick tests are more sensitive to albumin while Bence-Jones proteins, globulins, and glycoproteins are less readily detected. Dipstick is useful if urinary proteins are >300–500 mg/day or albumin >10–20 mg/dl.

8.2.2 The Boiling Test for Coagulable Proteins

When proteins are heated at pH equal to isoelectric point (pI), they get coagulated and denatured.

8.2.2.1 Reagents
33% (v/v) aqueous solution of acetic acid.

8.2.2.2 Procedure
Make the urine slightly acidic with acetic acid if urine is alkaline. Fill a test tube (3/4th) with acidified urine and heat the upper (1/3rd) portion until it boils. Compare it with unboiled urine lower down the tube. Appearance of faint turbidity to heavy precipitate indicates positive test. Phosphates also give precipitates, so to rule out false positive result, acetic acid is added. Phosphate get dissolved leaving any coagulated protein still visible.

Note Boiling test is semiquantitative test, so it can be used to measure concentration of proteins to some extent. No turbidity indicates negative result; only turbidity indicates ⍰30 mg/dl; turbidity with precipitates indicates 100 mg/dl, while turbidity with coagulation indicates proteins >300 mg/dl. If heavy coagulate appears, that indicates >2 g/dl proteins in urine sample.

8.2 Tests for Urinary Proteins

8.2.3 Sulphosalicylic Acid Test

Sulphosalicylic acid is an anionic protein precipitant; it reacts with protein cations and causes precipitation.

8.2.3.1 Reagent
Three percent aqueous solution of sulphosalicylic acid.

8.2.3.2 Procedure
Mix 1 ml of urine with 3 ml of sulphosalicylic acid. Presence of turbidity or white precipitate indicates proteins. Uric acid may give false positive test. To rule out this possibility, the sample is heated; if it becomes clear, proteins are absent and turbidity is due to uric acid.

8.2.4 Nitric Acid Ring Test (Heller's Test)

8.2.4.1 Reagents
Conc. HNO_3.

8.2.4.2 Procedure
Layer the urine carefully over few ml of conc. HNO_3 in a test tube so as to get a sharp line of demarcation. Proteins give a white color ring at the junction of fluids. However, urea, uric acid, and iodinated organic compounds used for X-rays of the urinary tract can give false positive test.

8.2.5 Bence-Jones Proteins

Bence-Jones proteins are light chains of immunoglobulins which precipitate at low temperature. The excretion of Bence-Jones proteins in urine is associated with multiple myeloma and malignant lymphoma. Bence-Jones proteins are synthesized by malignant plasma cells. There are three qualitative tests to indicate the presence of Bence-Jones proteins.

8.2.5.1 Harrison's Test

Principle
Bence-Jones proteins coagulate at temperature from 40 to 60°C, but the precipitated proteins dissolve at the boiling point (80–100°C). Subsequent cooling to 40–50°C reprecipitates the proteins.

Procedure
Add urine to 3/4 of test tube and add few drops of 33% acetic acid. The temperature is increased up to 40–60°C. Bence-Jones proteins get precipitated. If the temperature

is increased further, then precipitates get dissolved, and when cooled back to 40–60°C, the proteins precipitate again.

Precautions
1. Urine should be acidic and clear.
2. Temperature should be noted carefully.

8.2.5.2 Bradshaw's Test

Principle
Bence-Jones proteins get precipitated by HCl.

Procedure
Layer the urine carefully over conc. HCl in a test tube. If Bence-Jones proteins are present in urine, these will be precipitated by the HCl and form a fine or heavy ring at the interface of urine and HCl. Albumin may give a false positive test with HCl. To rule out the possibility of presence of albumin, dilute the urine with distilled water and repeat the test. The precipitate confirms the presence of Bence-Jones proteins.

8.2.5.3 Osgood and Haskin's Test

Principle
Bence-Jones proteins are precipitated in saturated sodium chloride (NaCl) solution at acidic pH.

Procedure
Add 0.5 ml of 50% acetic acid to 1.5 ml of saturated NaCl solution. To this, add 2.5 ml of fresh urine sample. The formation of precipitate at room temperature indicates the presence of Bence-Jones proteins.

Clinical Significance
The excretion of an abnormal protein amount in urine is a reliable marker of renal disease. When proteinuria is confirmed, 24 h urine sample is collected for protein analysis. This will indicate the degree of proteinuria. Depending on clinical finding, proteinuria can be divided into four types, viz., overload proteinuria, glomerular proteinuria, tubular proteinuria, and post renal proteinuria (Fig. 8.1).

Overflow or Overload Proteinuria Excessive production of proteinuria is due to hemoglobin or myoglobin and Bence-Jones proteins loss into urine. These proteins are not initially associated with glomerular or tubular disease but may cause renal disease. Myoglobin causes acute tubular necrosis. Bence-Jones proteins appear in urine in multiple myeloma.

Glomerular Proteinuria It is due to increased glomerular permeability. Glomerular disease often causes heavy proteinuria (excretion of proteins >3–4 g/day). Small

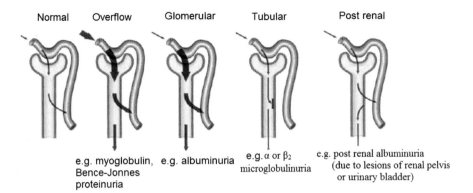

Fig. 8.1 Types of proteinuria

amount of albumin is observed in urine in insulin-dependent diabetics, and this finding appears to correlate with very early diabetic nephropathy. Immune complex disease also affects the glomerules. In such progressive renal disease, the ability to restrict filtration of smallest of larger proteins is lost first. Thus albumin appears first in urine. Large proteins are not seen in urine. Progressively, severe glomerular lesions produce less selective proteinuria that pass proteins of all sizes through the glomerulus and is called nonselective proteinuria. In the final stage of disease, as glomeruli are destroyed, proteinuria decreases and renal failure results.

Glomerular proteinuria with pathological damage to the glomerulus may be divided into two types, i.e., non-nephrotic proteinuria and nephrotic proteinuria based on the quantity of protein excretion. Nephrotic-range proteinuria is associated with loss of 3 g or more proteins in 24 h urine, while there is presence of 2 g of protein/g of urine creatinine in a single-spot urine collection. This finding denotes significant glomerular disease. Nephrotic syndrome is another clinical representation showing the combination of nephrotic-range proteinuria with a low serum albumin level and edema. The amount of proteinuria is <3.5 g/24 h and is persistent in case of non-nephritic proteinuria.

Tubular Proteinuria It is characterized by the appearance of low molecular weight proteins such as α-microglobulin, delta globulin such as $β_2$-microglobulin, light chain immunoglobulin, and lysozymes in the urine because of defective reabsorption of these compounds in the proximal renal tubules. The amount of proteinuria is about 1–2 g/day.

Post Renal Proteinuria It occurs due to proteins arising from the urinary tract and is usually due to inflammation or malignancy lesions (stones, tumor, growth) of renal pelvis, bladder, and prostate of urethra.

9

To Determine the Quantity of Proteins in Urine Sample Using Biuret Reaction

9.1 Theory

Healthy persons excrete urine that is largely free of proteins. The analysis of urine for the presence of proteins is important for the diagnosis of renal, cardiac, and thyroid-related diseases. Several physiological and pathophysiological mechanisms cause the urinary excretion of proteins. The normal protein excretion in urine is about 30–150 mg/day. About one third of the total urinary protein is albumin, and the remaining includes many small globulins. Small proteins are filtered readily in filtrate but are reabsorbed downstream by proximal tubular cells. The small amount of albumin filtered through the glomeruli is mostly absorbed by proximal tubular cells and degraded by lysosomal enzymes into fragments, which are returned to circulation in same way as that for small molecular weight proteins. The presence of excess proteins in urine often causes the urine to become foamy, but other conditions like presence of bilirubin in the urine (bilirubinuria), pneumaturia (air bubbles in the urine), or drugs may also make urine foamy. The presence of increased amounts of proteins in the urine may indicate a serious disease problem or may appear before any other clinical symptoms.

9.2 Specimen Requirements

A 12 h or 24 h urine specimen without any preservative is preferred. Use fresh sample otherwise store at 2–8°C for up to 48 h.

9.3 Principle

The protein in urine is precipitated with trichloroacetic acid (TCA), redissolved in alkali, and measured colorimetrically using the Biuret reaction. Biuret reagent contains alkaline $CuSO_4$ solution and sodium-potassium tartrate. The Cu^{2+} ions form a coordination complex with four imino groups present in the peptide bonds and show an absorption maximum at 540 nm. The method is suitable for estimation of proteins having concentration greater than 1 mg. The intensity of color produced is directly proportional to the amount of protein present in the sample.

9.4 Reagents

1. **20% TCA** – Dissolve 20 g TCA in final volume of 100 ml distilled water.
2. **0.5 N NaOH** – Dissolve 2 g of NaOH in final volume of 100 ml distilled water.
3. **Bovine serum albumin (BSA) standard (0.5 g/dl)** – Dissolve 0.5 g BSA in 100 ml distilled water or normal saline. Do not shake.
4. **Biuret reagent** – Dissolve 9 g of sodium-potassium tartrate in 500 ml of 0.2 N NaOH solution. To this add 3 g of $CuSO_4$ previously dissolved in 100 ml distilled water, gently with constant mixing. Then dissolve 5 g potassium iodide and make final volume up to 1 litre with 0.2 N NaOH.

Procedure

1. Take a set of test tubes in duplicate and label as blank, standard (S_1–S_5) and test (T).
2. Add BSA standard and distilled water as shown in table.
3. Then add 20% TCA to all tubes except blank.
4. Mix and incubate all the tubes at 37°C for 30 min.
5. Centrifuge all the test tubes at 5000 rpm for 10–15 min and, discard the supernatant completely.
6. Dissolve the pellet in 0.5 ml NaOH and add 1.5 ml distilled water to all tubes.
7. Then add 2 ml biuret reagent to all the tubes, incubate at 37°C for 10 min, and take absorbance at 540 nm.

Reagent	Blank	S_1	S_2	S_3	S_4	S_5	T
BSA (ml)	–	0.2	0.4	0.6	0.8	1.0	–
BSA amount[a] (mg)	–	1	2	3	4	5	–
Urine (ml)	–	–	–	–	–	–	0.2

(continued)

9.6 Clinical Significance

Distilled water (ml)	1	0.8	0.6	0.4	0.2	–	0.8
20% TCA (ml)	–	1	1	1	1	1	1
Mix and incubate all the tubes at 37°C for 30 min. Centrifuge all the test tubes at 5000 rpm for 10–15 min and discard the supernatant completely.							
0.5 N NaOH (ml)	0.5	0.5	0.5	0.5	0.5	0.5	0.5
Distilled water (ml)	1.5	1.5	1.5	1.5	1.5	1.5	1.5
Biuret reagent (ml)	2	2	2	2	2	2	2

[a]Amount present in volume of BSA standard added in test tubes

9.5 Calculations

Plot a graph between amount of BSA at x-axis and absorbance at y-axis. Then extrapolate quantity of proteins present in given urine sample from the graph. The extrapolated amount will give amount of protein in 0.2 ml of urine sample. Express protein amount per liter of urine sample or 24 h urine sample. The protein concentration in urine sample can also be calculated by using equation:

$$\text{Amount of protein} = \frac{\text{OD of test} \times \text{amount of standard(mg)} \times 1000}{\text{OD of standard} \times \text{volume of sample(ml)} \times 1000}$$
$$= x \text{ g/L of urine}$$

Note For calculation, take BSA protein amount and OD of same standard, i.e., if amount of S_2 standard is taken, then take OD of S_2 only. In formula, the value 1000 at denominator is taken to convert protein amount to per gram and at numerator is used to convert value to per liter of urine.

Precautions

1. Filter or centrifuge the urine, if urine sample is turbid.
2. Don't dissolve $CuSO_4$ in sodium-potassium tartrate solution directly in the preparation of Biuret reagent.

9.6 Clinical Significance

The clinical significance of proteinuria has been discussed in the previous chapter.

To Estimate the Amount of Total Protein and Albumin in Serum and to Find A/G Ratio

10

10.1 Theory

Proteins are present in all body fluids but show very high concentration (> 3 g/dl) in plasma, lymphatic fluids, and some exudates. The amount of total proteins in serum decreases in the third trimester of pregnancy. The measurement of total proteins in serum is useful to assess the conditions related to changes in plasma or fluid volumes, such as shock and dehydration. A total serum protein test measures the total amount of proteins in the blood. Serum proteins mainly consist of albumin but few globulins (such as α_1-globulin) also may be measured. Fibrinogen is the protein present in plasma but not in serum. Albumin is a major constituent of human plasma and represents about 40–60% of total proteins. After its synthesis in the liver (hepatocytes), it is secreted in plasma, a process dependent on protein intake. The half-life of albumin is 15–19 days. Albumin has molecular weight of approximately 66,000 dalton which makes it a marker of both glomerular and blood-brain barrier functions. The pI of albumin lies between 4 and 5.8; hence it is present as an anion at pH 7.4. Albumin is involved in transport and storage of wide variety of ligands and maintaining the osmotic pressure of plasma. It also serves as a source of endogenous amino acids. Albumin is mainly measured to check liver and kidney functions but is also important in determining the cause of swelling of the ankles (edema) or abdomen (ascites). The globulins constitute various proteins like α, β, and γ types that can be separated into distinct bands on electrophoresis. Few globulins are synthesized by the liver while others by the immune system. Globulins perform wide variety of functions including transport and immunity.

10.2 Principle

When a protein solution is treated with Cu^{2+} in an alkaline solution, a colored chelate complex of unknown composition is formed between Cu^{2+} and carbon ion and imine group of peptide bond. An analogous reaction takes place between Cu^{2+} and organic compound biuret, and therefore the reaction is called the biuret reaction. Amino acids and dipeptides do not give this reaction, but tri- and polypeptides can react. In biuret reaction, one copper ion is linked to 4–6 nearby peptide linkages by coordinate bonds. The intensity of color produced is proportional to the number of peptide bonds undergoing reaction.

10.3 Specimen Requirements

Both serum and plasma may be used for protein estimation, but serum is preferred over plasma. A fasting specimen may be desired to decrease lipemia since hemolyzed and lipemic samples strongly interfere with the protein measurement by biuret method. Samples remain stable for 24 h if stored at room temperature.

10.4 Reagent

1. **Biuret reagent** – Dissolve 9 g of sodium-potassium tartrate in 500 ml of 0.2 N NaOH solution. To this add 3 g of $CuSO_4$ previously dissolved in 100 ml distilled water, gently with constant mixing. Then dissolve 5 g of potassium iodide, and make final volume up to 1 liter with 0.2 N NaOH.
   Sodium-potassium tartrate is used as a complexing agent to keep the copper in Cu^{2+} state in the solution. Potassium iodide is added to prevent auto-reduction. NaOH provides alkaline medium.
2. **Biuret blank**: Prepare same as biuret reagent with all reagents except $CuSO_4$.
3. **Protein standard (10% BSA)**: Dissolve 10 g BSA in 100 ml normal saline.

4. **Working protein standard:** Dilute stock standard with normal saline to get concentration range of 2, 4, 6, 8, and 10 g/dl.

10.5 Procedure

1. Take a set of test tubes in duplicate and label as blank, standard (S_1–S_5), and test (T).
2. Add 0.1 ml protein standard of concentration 2, 4, 6, 8, and 10 gm/dl, respectively, in all tubes labelled as S_1–S_5.
3. In tube T add 0.1 ml serum sample.
4. Then add 5 ml biuret reagent to all tubes except blank. In blank, add 5 ml biuret blank.
5. Mix and incubate all the tubes at 37°C for 30 min and take OD at 540 nm.

Reagents	Blank	S_1	S_2	S_3	S_4	S_5	T
BSA (ml)	–	0.1	0.1	0.1	0.1	0.1	–
BSA concentration (g/dl)	–	2	4	6	8	10	
BSA amount (mg)	–	2	4	6	8	10	–
Serum (ml)	–	–	–	–	–	–	0.1
Distilled water (ml)	0.1	–	–	–	–	–	–
Biuret reagent (ml)	–	5	5	5	5	5	5
Biuret blank (ml)	5	–	–	–	–	–	–

10.6 Calculations

Plot a graph between amount of BSA at x-axis and absorbance at y-axis. Then extrapolate quantity of protein in the given serum sample from the graph. The amount extrapolated will denote protein amount in 0.1 ml of sample. Calculate and express protein amount per 100 ml serum sample. The protein concentration in serum sample can also be calculated by using the following equation:

$$\text{Amount of protein in serum} = \frac{\text{OD of test} \times \text{amount of standard (mg)} \times 100}{\text{OD of standard} \times \text{volume of sample (ml)} \times 1000^*}$$
$$= x \, gm/dl$$

*To convert mg to gram

10.7 Estimation of Albumin in Serum

Method Bromocresol Green (BCG) Binding

10.8 Principle

Many colored anionic dyes such as methyl orange and BCG bind with albumin. The use of BCG is free from pigment interference. Albumin at a pH below its pI is positively charged and has affinity for anionic dye (BCG) which on combining with it undergoes a change in color from yellow to blue-green which is measured at 620 nm. The color produced is directly proportional to albumin concentration.

10.9 Reagents

1. **BCG solution**
   Sol A: Succinate buffer (0.5 M, pH 4.1): Dissolve 0.56 g succinic acid in approximately 80 ml distilled water. Adjust pH to 4.1 with 1 N NaOH, and make final volume up to 100 ml with distilled water and store at 4°C.
   Sol B: BCG dye (10 mM): Dissolve 10 mg BCG dye in 1 ml of 1 N NaOH, and make the final volume 10 ml with distilled water.
   Sol C: Sodium azide solution: Dissolve 4 g sodium azide in distilled water, and make final volume up to 100 ml.
   Sol D: Brij 35: Prepare 30% in distilled water.
   Prepare BCG solution by mixing 50 ml of solution A, 4 ml of solution B, 1.25 ml of solution C, and 1.25 ml of solution D. Make the volume to 500 ml with distilled water, and adjust the pH to 4.1 ± 0.05 by adding 1 N NaOH.
   Sodium azide solution is added as a preservative. Brij 35 is added to prevent the turbidity produced when sample is mixed with reagent. It also causes a slight change in absorbance maxima of dye-albumin complex.
2. **Stock BSA standard**: 5 g/dl in normal saline.
3. **Working BSA standard** of concentration 1, 2, 3, 4, and 5 g/dl by diluting the stock BSA standard in normal saline.
4. **0.9% aqueous solution of normal saline**.

10.10 Procedure

1. Take a set of seven test tubes in duplicate and label as blank, standard (S_1–S_5), and test (T).
2. Add 1 ml of normal saline in all the tubes.
3. Add 0.02 ml of working standard of concentration 1, 2, 3, 4, and 5 g/dl in respective test tubes.

4. In tube T, add 0.02 ml of serum sample and in blank add 0.02 ml of distilled water.
5. Then add 4 ml of BCG solution in all the tubes and immediately read at 620 nm.

Reagents	Blank	S_1	S_2	S_3	S_4	S_5	T
Normal saline (ml)	1	1	1	1	1	1	1
BSA (ml)	–	0.02	0.02	0.02	0.02	0.02	–
BSA concentration (g/dl)	–	1	2	3	4	5	–
BSA amount (mg)	–	1	2	3	4	5	–
Serum (ml)	–	–	–	–	–	–	0.02
Distilled water (ml)	0.02	–	–	–	–	–	–
BCG Bromo cresol green (BCG) solution (ml)	4	4	4	4	4	4	4

10.11 Calculations

Plot a graph between amount of BSA at x-axis and absorbance at y-axis. Then extrapolate quantity of albumin in given serum sample from the graph. The albumin concentration in serum sample can also be calculated by using the following equation:

$$\text{Amount of albumin in serum} = \frac{\text{OD of test} \times \text{amount of standard (mg)} \times 100}{\text{OD of standard} \times \text{volume of sample (ml)} \times 1000}$$
$$= x \text{ g/dl}$$

10.12 Clinical Significance

The normal levels of total proteins in serum are 5.5–8.0 g/dl. The total protein concentration may be altered by changes in plasma volume. Increase in protein concentration may occur in dehydration, due to increased globulin synthesis, acute or chronic infection, and multiple myeloma. The decrease in protein levels occurs due to low protein intake, starvation, malnutrition, and decreased synthesis due to liver diseases or nephrotic syndrome.

The normal serum albumin value is 3.5–5.5 g/dl and globulin is 2.0–3.5 g/dl. The albumin/globulin (A/G) ratio is 1.5:2.5 (serum globulin levels are measured by subtracting serum albumin levels from total serum proteins). In a very low or high A/G ratio, further testing must be performed to determine the cause and diagnosis. A low A/G ratio may indicate autoimmune disease, multiple myeloma, cirrhosis, or kidney disease, while high A/G ratio suggests genetic deficiencies or leukemia.

Albumin amount increases in dehydration state, the condition is counteracted in vomiting and diarrhea, and this may impair intestinal absorption of amino acids, so, reducing plasma-albumin formation. Albumin concentration decreases due to loss in urine in nephrotic syndrome, chronic glomerular nephritis, diabetes, and severe hemorrhage. Synthesis of albumin is reduced in liver disease, malnutrition, defective digestion or malabsorption, and increased catabolism of proteins in the liver. Albumin levels generally parallel total protein levels, except when total protein changes due to gamma globulins.

10.13 Precautions

1. pH of BCG solution should be carefully adjusted and should be exactly 4.1 ± 0.05.
2. Reading should be taken immediately after adding BCG solution to avoid interference of globulin.

To Perform Qualitative Test for Reducing Substances in Urine

11

11.1 Theory

Sugars may be called as reducing or nonreducing based on their ability to reduce copper during the Benedict's test. The reducing property of sugar is based on the presence of free aldehyde or ketone group in them. Most of monosaccharides and disaccharides are reducing sugars, while sucrose is nonreducing sugar. Reducing sugars are capable of reducing Cu^{2+} (cupric ions) to Cu^+ (cuprous ions) in alkaline medium which produces red precipitate of cuprous oxide or yellow precipitate of cuprous hydroxide. The free carbonyl carbon of aldehyde or keto sugars shows the reducing property. The urine of normal individuals contains small amount of reducing substances which are not sufficient to give positive test with Benedict's test or Fehling's test. Various reducing sugars present in the urine are glucose, galactose, fructose, and lactose.

11.2 Qualitative Test for Reducing Sugars

11.2.1 Benedict's Test

Principle
Benedict's qualitative reagent contains Cu^{2+} ions complexed with citrate in alkaline medium. The free aldehyde or keto group of the reducing carbohydrates reduces cupric ions to cuprous ion with the resultant formation of yellow or red precipitates of cuprous oxide. Sodium citrate prevents the spontaneous reduction of $CuSO_4$ while Na_2CO_3 is used to provide alkaline medium.

$$CuSO_4 + 2H_2O \xrightarrow{sugar} Cu(OH)_2 + H_2SO_4$$
Blue → white

$$2Cu(OH)_2 \xrightarrow{heat} Cu_2O + 2H_2O + O$$
Cuprous oxide
(red or yellow)

11.2.2 Benedict's Qualitative Reagent

Sol A: Dissolve 17.3 g of sodium citrate and 10 g of Na_2CO_3 in about 60 ml of distilled water with the aid of heat.
Sol B: Dissolve 1.73 g of $CuSO_4$ in 10 ml of distilled water.

Solution B is added to solution A gently with constant mixing, and final volume is made up to 100 ml with distilled water.

11.2.3 Procedure

Take 5 ml Benedict's reagent in a test tube and add 0.5 ml or eight drops of urine. Mix and heat it to boil for 3–5 min and allow cooling. Observe the color, and report the result on the basis of color produced as given in table below.

S. no	Approximate amount of sugar	Appearance of color change	Result
1	Nil	Blue	Nil
2	Traces	Bluish green	+
3	0.5 g/dl	Green precipitate	+
4	1 g/dl	Yellow precipitate	++
5	1.5 g/dl	Brown precipitate	+++
6	>2 g/dl	Brick red precipitate	++++

Sensitivity of the test is 0.1–1.5% toward false-positive results obtained from other reducing substances such as lactose, galactose, fructose, xylose, etc. Specific tests are performed to rule out the possibility of presence of these sugars. Normal constituents such as uric acid, creatinine, and ascorbic acid in higher amount also give false-positive result.

11.3 CLINISTIX/Uristix

CLINISTIX are commercially available strips. These are impregnated with glucose oxidase (GOD), peroxidase (POD), and dye O-toluidine (chromogen). In this reaction, glucose is oxidized to gluconic acid and H_2O_2 by the enzyme GOD. H_2O_2 formed is split into water and nascent oxygen by POD enzyme. This nascent oxygen reacts with the chromogen to form a colored product. Sensitivity of the test is 0.01–1 g/dl.

$$\text{Glucose} \xrightarrow{\text{GOD}} \text{Gluconic acid} + H_2O_2$$
$$H_2O_2 \xrightarrow{\text{POD}} H_2O + [O]$$
$$[O] + o-\text{toluidine} \rightarrow \text{coloured complex}$$

11.3.1 Procedure

Strip is dipped in urine sample for 10 sec and color is observed after 30 min. Compare the color formed with the chart provided.

11.4 Precautions

1. The container used for urine collection should be clean and free from contaminants, particularly disinfectants and detergents containing oxidizing substances such as peroxides.
2. Do not touch the test area of the strip.

11.5 Clinical Significance

Reducing sugars are found only in very small amounts in the urine of normal persons, usually less than 100 mg/day. Amount of sugars increases under some pathological conditions that can be detected by various tests.

1. Glucose: Glucose is the reducing sugar found in urine which is pathologically significant. Normally less than 100 mg of glucose (sugar) is excreted per day. Glucose is present in urine whenever blood glucose rises above the renal threshold, i.e., 180 mg/dl blood. Excretion of glucose in urine is called glycosuria. Glycosuria is found in all conditions in which glucose tolerance is diminished. Persistent glycosuria may indicate diabetes mellitus. In diabetes mellitus, 5–6% glucose is excreted through urine of patient with endocrine hyperactivity. Severe liver disease and whole organ disease of the pancreas may also be accompanied by some glycosuria.
2. Lactose: It is found in the urine of lactating woman and toward the later stage of pregnancy.
3. Galactose: It is present in urine very rarely. It occurs occasionally in lactation. Galactosuria occurs in infants in recessive inherited disorder and is due to an inability to metabolize galactose derived from lactose in the milk.

4. Fructose: It may be found in urine after taking food rich in fructose (fruits, honey, syrup, and jams). It may be found in liver disease and in the urine of diabetics along with glucose.

Apart from these sugars, pentoses, glucuronides, and homogentisic acid are also present under some pathological conditions which can also reduce Benedict's reagent. Higher concentration of uric acid and creatinine also reduce Benedict's qualitative reagent.

Quantitative Analysis of Reducing Sugars in Urine

12.1 Theory

Examination of urine for glucose is rapid and inexpensive and can be used to screen large number of samples. The screening test detects sugars that reduce copper and produce different shades of colors. Substances other than glucose also reduce $CuSO_4$ and produce positive result. Quantitative test estimates the amount of reducing substances present in urine. This amount is sufficiently small to be detected positively with qualitative methods of glucose determination. The condition characterized by excretion of glucose in urine is called glycosuria. The commonly used quantitative test for urinary sugars is titration based Benedict's test. This method is not specific for glucose and measures all the reducing sugars present in urine sample.

12.2 Specimen Requirement

In quantitative analysis, 24 h urine specimen is used and is preserved by adding 5 ml glacial acetic acid (to inhibit the bacterial growth) to the container before starting the collection. Urine should be stored at 4°C during collection.

12.3 Principle

Glucose or any other reducing carbohydrate reduces Benedict's quantitative reagent to a white precipitate of cuprous thiocyanate (instead of usual reddish brown cuprous oxide). Besides the usual ingredient of Benedict's qualitative reagent, quantitative reagent contains potassium thiocyanate and potassium ferrocyanide. The disappearance of the blue color and appearance of white precipitate indicates complete reduction of copper, which is much easier to detect. Potassium ferrocyanide prevents the precipitation of cuprous oxide and keeps the cuprous ions in solution. Na_2CO_3 prevents destruction of sugar by providing the supply of OH ions.

12.4 Reaction

$$\text{Potassium thiocyanate} + Cu^{2+} \text{ (blue)} \rightarrow \text{cuprous thiocynate (white ppt)}$$

Addition of Na_2CO_3 also helps to visualize end point, which is the disappearance of blue color.

12.5 Reagents

1. *Benedict's quantitative reagents:*
   Sol. A: Dissolve 20 g of sodium citrate, 7.5 g of Na_2CO_3, and 1.25 g of potassium thiocyanate in about 60 ml warm water, and cool the solution.
   Sol. B: Dissolve 1.8 g of $CuSO_4$ in 10 ml of distilled water.

Add solution B to solution A gently with constant mixing. To the resultant solution, add 5% aqueous solution of potassium ferrocyanide. Make the final volume to 1 litre with distilled water.

2. *Na_2CO_3 anhydrous salt*

12.6 Procedure

1. Pipette out 25 ml of Benedict's quantitative reagent in 100 ml conical flask. To this add 3–4 g of Na_2CO_3.
2. Add few glass beads to reduce bumping and bring to a boil.
3. While still in boiling condition, add urine slowly dropwise using the burette, and mix the contents by shaking.
4. Add the urine until the blue color changes to white.
5. Note the volume of urine used. Repeat the titration to get concordant values.

12.7 Precautions

1. Add marbles or glass beads to avoid bumping.
2. Sample should be added to boiling reagent slowly with constant shaking.

3. If the urine sample used is less than 2 ml for the titration, dilute urine 1 in 5 or 1 in 10 with distilled water and include the dilution factor in the calculation, after doing a rough titration.
4. A greenish-blue color will reappear if flask is allowed to stand. This is reoxidation process of air and may be ignored.

12.8 Calculation

50 mg of glucose will reduce 25 ml of Benedict's quantitative reagent. Suppose titer of urine is 10 ml, then 10 ml urine must contain 0.05 g glucose and then amount of glucose/100 ml urine

$$= \frac{50 \times 100}{10}$$
$$= x\,mg/dl$$

12.9 Clinical Significance

The clinical significance of glycosuria has been discussed in previous chapter.

Estimation of Blood Glucose Levels by Glucose Oxidase Method

13.1 Theory

The sugar glucose is an aldohexose and is the most important carbohydrate present in our body. The bulk of dietary carbohydrate is absorbed mainly as glucose in the intestine. Glucose is the major source of energy for all the tissues including the brain. It enters the cell through the influence of insulin and undergoes a series of chemical reactions to produce energy. It acts as precursor for the synthesis of all other carbohydrates in the body which have highly specific functions, e.g., glycogen for storage, ribose in nucleic acid, and galactose in lactose of milk. High levels of glucose are observed in diabetes mellitus, while decrease in glucose levels occurs in starvation and hyperinsulinemia.

13.2 Principle

The enzyme GOD oxidizes the plasma glucose to gluconic acid with the liberation of H_2O_2, which is converted to water and oxygen by the enzyme POD. 4-aminoantipyrine, an oxygen acceptor, takes up the oxygen and together with phenol forms a pink-colored product which can be measured at 520 nm.

$$\text{Glucose} \xrightarrow{\text{GOD}} \text{Gluconic acid} + H_2O_2$$
$$H_2O_2 \xrightarrow{\text{POD}} H_2O + [O]$$
$$[O] + \text{4-aminoantipyrine} + \text{Phenol} \rightarrow \text{Chromogen (coloured)}$$

13.3 Specimen Requirements

Plasma is the preferred sample for glucose estimation. Separate plasma from cells rapidly to avoid glucose loss. It is stable for 8 h at 25°C and 72 h at 4°C. Preserve plasma with sodium fluoride. Whole blood may also be used, but whole blood levels are 90% of total plasma glucose.

13.4 Reagents

1. **Phosphate buffer**: Prepare 0.1 M potassium phosphate buffer of pH 7.0.
2. **Enzyme reagent**: Dissolve 18 mg of 4-aminoantipyrine and 36 mg solid phenol in 100 ml of 0.1 M phosphate buffer. Add GOD (1500 U) and POD (100 U). Store in brown bottle at 4°C.
3. **Stock glucose solution (100 mg/dl)**: Dissolve 100 mg glucose in 100 ml distilled water.
4. **Working glucose standard**: Prepare glucose standards of concentration 25, 50, 75, and 100 mg/dl.

13.5 Procedure

1. Take a set of test tubes in duplicates and label as B, S_1, S_2, S_3, S_4, S_5, and T.
2. Add 1.8 ml phosphate buffer and 1 ml enzyme reagent to all the test tubes. In blank add 1.9 ml phosphate buffer and 1 ml enzyme reagent.
3. Add 0.1 ml of standard of concentration 10, 25, 50, 75, and 100 mg/dl to the tubes labelled as S_1, S_2, S_3, S_4, and S_5, respectively. Add 0.1 ml of serum sample in tube T.
4. Mixed the content of the tubes thoroughly and place them at room temperature for 15 min.

Reagents	Blank	S_1	S_2	S_3	S_4	S_5	T
Phosphate buffer (ml)	1.9	1.8	1.8	1.8	1.8	1.8	1.8
Enzyme reagent	1.0	1.0	1.0	1.0	1.0	1.0	1.0
Glucose standard (ml)	–	0.1	0.1	0.1	0.1	0.1	–
Glucose concentration (mg/dl)	–	10	25	50	75	100	–
Glucose amount (mg)		0.01	0.25	0.5	0.75	1.0	
Serum/plasma (ml)	–	–	–	–	–	–	0.1
Mix well and keep at room temperature for 15 min and take OD at 520 nm							

13.6 Calculations

Plot a graph between amount of glucose at x-axis and absorbance at y-axis. Extrapolate glucose amount from the given sample from the graph. The glucose concentration in serum sample can also be calculated by using the following equation:

$$\text{Plasma glucose} = \frac{\text{OD of test} \times \text{amount of standard (mg)} \times 100}{\text{OD of standard} \times \text{volume of sample (ml)}}$$
$$= x \, \text{mg/dl}$$

13.7 Clinical Significance

Normal range of blood glucose is 70–110 mg/dl. Increase in blood glucose in disease conditions causes hyperglycemia, and decrease in blood glucose causes hypoglycemia. Hyperglycemia is associated with diabetes mellitus and hyperactivity of thyroid, pituitary, and adrenal glands. A moderate hyperglycemia may also be found in some intracranial disease such as meningitis, encephalitis, tumors, and hemorrhage. Depending upon the type, degree, and duration of the anesthesia, quite a considerable rise in blood glucose may occur, sometimes to over 200 mg/dl, thus producing a temporary glycosuria. The glucose concentration in fasting samples of 150–200 mg/dl is very suggestive of diabetes mellitus, and over 200 mg/dl is almost diagnostic. For fasting samples, a 6–8 h fast is required. Hyperglycemia is also observed in pancreatitis and carcinoma of pancreas where increase in fasting blood sugar may occur but except in advanced cases; levels are moderate and does not exceed 150 mg/dl.

Hypoglycemia occurs most frequently as a result of overdosage with insulin in the treatment of diabetes. Insulin secreting tumors of pancreas produce severe hypoglycemia. The fasting blood glucose may be reduced in hypothyroidism (myxedema, cretinism) and hypoadrenalism (Addison's disease), and starvation and severe exercise may produce hypoglycemia. Increased glycogen storage in the liver may cause hypoglycemia due to deficiency of an enzyme involved in the breakdown of glycogen to glucose. Impaired absorption of glucose from intestine in some type of steatorrhea also causes hypoglycemia, while alcohol ingestion also causes hypoglycemia, due to increased oxidation of ethanol by forming more NADH and reducing the available NAD^+ (Fig. 13.1).

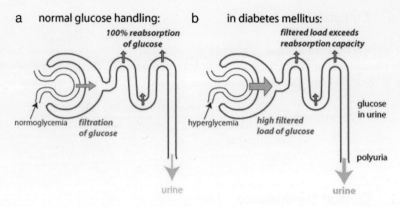

Fig. 13.1 (a) Glucose reabsorption in normal conditions and (b) glucose excretion in diabetes

To Determine the Blood Glucose Levels by Folin and Wu Method

14.1 Principle

The test is based on reducing property of glucose in hot alkaline solution. Proteins are precipitated by tungstic acid and removed by centrifugation. Protein-free filtrate of plasma contains glucose that reduces cupric ions of alkaline $CuSO_4$ to cuprous oxide, which in turn reduces phosphomolybdic acid to molybdenum blue, and color thus produced is measured colorimetrically at 430 nm.

14.2 Reagents

1. **N/12 H_2SO_4 acid**: 0.25 ml conc. H_2SO_4 is diluted to 100 ml with distilled water.
2. **Sodium tungstate (10%)**: Dissolve 1 g sodium tungstate in distilled water to make final volume of 10 ml.
3. **Alkaline copper sulfate reagent**
   Sol. A – Dissolve 0.9 g of $CuSO_4.5H_2O$ in 50 ml of distilled water.
   Sol. B – Dissolve 0.8 g of Na_2CO_3 and 0.15 g of tartaric acid in 50 ml distilled water. Mix solution A and B in equal volume to form solution of alkaline $CuSO_4$.
4. **Phosphomolybdic acid**: Dissolve 3.5 g of phosphomolybdic acid and 0.5 g of sodium tungstate and 2.0 g of NaOH in about 20 ml of distilled water. Boil for 30–40 min for removal of ammonia. Cool and add slowly 1.25 ml of phosphoric acid (H_3PO_4) with constant stirring, and make the final volume to 50 ml with distilled water.
5. **Glucose standard stock solution (500 mg/dl)**: Dissolve 500 mg of glucose in saturated benzoic acid solution (0.2%) and make up to 100 ml with it.
6. **Working glucose standards**: Prepare 25, 50, 75, 100, 200, 400, and 500 mg % glucose solutions by diluting the stock glucose standard with saturated benzoic acid.

14.3 Procedure

Take nine test tubes in duplicate, and label them as blank, S_1–S_7 (standard) and T (test). Add standard glucose solution of 25, 50, 75, 100, 200, 300, and 500 mg % in test tubes S_1–S_7, respectively. Then add reagents according to protocol given in the table. Read absorbance at 430 nm.

Reagents	Blank	S_1	S_2	S_3	S_4	S_5	S_6	S_7	T
N/12 H_2SO_4 (ml)	1.6	1.6	1.6	1.6	1.6	1.6	1.6	1.6	1.6
Glucose standard (ml)	–	0.2	0.2	0.2	0.2	0.2	0.2	0.2	–
Glucose concentration (mg/dl)	–	25	50	75	100	200	300	500	–
Glucose amount (mg)	–	0.025	0.05	0.075	0.1	0.2	0.3	0.5	
Distilled water (ml)	0.2	–	–	–	–	–	–	–	–
Serum/plasma (ml)	–	–	–	–	–	–	–	–	0.2
Sodium tungstate (ml)	0.2	0.2	0.2	0.2	0.2	0.2	0.2	0.2	0.2
Mix well and wait for 10 min, and then centrifuge at 3000 rpm for 10 min. Collect supernatant									
Supernatant (ml)	0.5	0.5	0.5	0.5	0.5	0.5	0.5	0.5	0.5
Alkaline $CuSO_4$ (ml)	1.0	1.0	1.0	1.0	1.0	1.0	1.0	1.0	1.0
Mix well and keep in boiling water bath for 10 min and cool									
Phosphomolybdate (ml)	1.0	1.0	1.0	1.0	1.0	1.0	1.0	1.0	1.0
Shake well to remove CO_2 completely									
Distilled water (ml)	2.5	2.5	2.5	2.5	2.5	2.5	2.5	2.5	2.5

14.4 Calculations

Plot a graph between amount of glucose at x-axis and absorbance at y-axis. Extrapolate glucose amount in the given sample from the graph or calculate using equation as explained in previous experiment.

To Perform Glucose Tolerance Test

The ability of a person to metabolize a given glucose is called glucose tolerance. To assess carbohydrate metabolism status, an oral glucose test is employed. Glucose tolerance test (GTT) is highly useful to confirm diabetes mellitus.

15.1 Method of Carrying GTT

The test is usually carried out in the morning. Patient should be on 8–12 h fasting before undergoing test and should not eat during test. Patient is given glucose solution (75 g) and is asked to drink it within 15 min. For children 1.75 g glucose/kg body weight is given. Blood for estimation of glucose is taken at half-hourly intervals for 2.5 h after the glucose intake. Fasting and every half-hourly urine specimens are also collected from the patient (six urine samples). Blood glucose concentration is estimated by GOD method. Urine sugar is detected by Benedict's qualitative test.

15.2 Specimen Requirements

Serum/plasma sample is used. For oral GTT, draw fasting sample and then at 30, 60, 90, and 120 min after glucose intake. In case of pregnancy, draw samples at fasting and at 60, 120, and 180 min after glucose intake.

15.3 Principle

The estimation of plasma glucose is based on the principle as explained for glucose measurement by glucose oxidase method.

15.4 Reagents

1. **Phosphate buffer (0.1 M, pH 7.0)**.
2. **Enzyme reagent**: Prepare as explained in previous experiment.
3. **Stock glucose solution (500 mg/dl)**: Dissolve 500 mg glucose in 100 ml distilled water.
4. **Working glucose standard**: Prepare glucose standards of concentration 50, 100, 150, 200, and 250 mg/dl.

15.5 Procedure

1. Take set of 11 test tubes in duplicate and label as blank, standard (S_1–S_5), and test samples (T_1–T_5).
2. Add 1.8 ml phosphate buffer to all the test tubes. In blank add 1.9 ml phosphate buffer.
3. Add 1 ml enzyme reagent to all the test tubes.
4. Add 0.1 ml of standards of concentration 50, 100, 150, 200, and 250 mg/dl to the tubes labelled as S_1–S_5, respectively.
5. Add 0.1 ml of serum sample to test tubes (T_1–T_5). Mix the content of the tubes thoroughly, and place them in a water bath at 37°C for 5 min or at room temperature for 15 min.
6. Measure the OD of the test and the standards against blank at 520 nm.

15.5 Procedure

Reagents	Blank	S_1	S_2	S_3	S_4	S_5	T_1 (fasting)	T_2 (30 min)	T_3 (60 min)	T_4 (90 min)	T_5 (120 min)	T_2 (150 min)
Phosphate buffer (ml)	1.9	1.8	1.8	1.8	1.8	1.8	1.8	1.8	1.8	1.8	1.8	1.8
Working reagent (ml)	1	1	1	1	1	1	1	1	1	1	1	1
Glucose standard (ml)	–	0.1	0.1	0.1	0.1	0.1	–	–	–	–	–	–
Glucose amount (mg)	–	0.05	0.1	0.15	0.2	0.25						
Serum/plasma (ml)	–	–	–	–	–	–	0.1	0.1	0.1	0.1	0.1	0.1
Mix well and keep at 37°C for 5 min and read at 520 nm												

15.6 Calculations

Plot a graph between amount of glucose at x-axis and absorbance at y-axis. Extrapolate glucose amount of the given sample from the graph or calculate using the equation below:

$$\text{Plasma glucose} = \frac{\text{OD of test} \times \text{amount of standard (mg)} \times 100}{\text{OD of standard} \times \text{volume of sample (ml)}}$$
$$= x \, \text{mg/dl}$$

15.7 Clinical Significance

In normal persons, the fasting blood glucose level is within 70–110 mg/dl. Following oral intake of glucose, the level rises and reaches at peak within 1 h and then come to normal fasting levels within 1.5–2 h because of insulin secretion and utilization and storage of glucose. After 2 h it should be below 120 mg/dl. There should be a negative test for glucose in all urine samples. In diabetes mellitus, the glucose tolerance is decreased. The glucose level increases to much higher levels after 1 h of glucose intake and does not reach normal level even after 2 h due to low insulin level or decreased cellular response toward insulin. In impaired glucose tolerance, the blood sugar values are below diabetic level but above normal (Fig. 15.1). In renal glycosuria, the glucose tolerance is normal; however glucose will appear in urine.

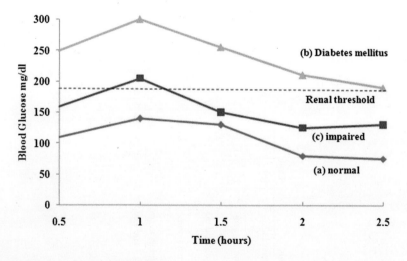

Fig. 15.1 Glucose tolerance test curve

Estimation of Urea in Serum and Urine 16

16.1 Theory

Urea is the major nitrogen-containing metabolite product of protein catabolism in humans, accounting for more than 75% of the nonprotein nitrogen eventually excreted. The biosynthesis of urea is carried out by hepatic enzymes of urea cycle. It is formed in the liver from carbon dioxide and ammonia, passes into the extracellular fluid, and is excreted almost entirely by the kidneys. The measurement of urea is an important investigation in diagnosing kidney damage. Urea is released into the blood which is then cleared by kidneys. The normal range of blood urea is 15–40 mg/dl. It is higher in men than women and more in adults than young. The urea content over periods is influenced by the amount of protein in the diet and tends to be lower in people on low protein diet. On an ordinary diet, urea nitrogen forms about 80–90% of the total urine nitrogen, but on low protein diet, it falls toward 60%. The total daily excretion of urea is about 30–40 g. Increased urea production occurs on high protein diets or after gastrointestinal hemorrhage and when there is increased tissue breakdown as observed in starvation, trauma, and inflammation. The capacity of the normal kidney to excrete urea is high, and in the presence of normal renal functions, urea levels rarely rise above normal despite increased production. A plasma urea concentration above 15 mmol/L almost certainly indicates renal impairment. The plasma urea is the most useful test of "renal excretory function," as it correlates well with the clinical consequences of retained metabolic products (uremia) in renal insufficiency.

16.2 Specimen Requirements

Serum/plasma or 24 h urine sample with preservative is used. Avoid high concentration of sodium fluoride during enzymatic estimation in serum. Serum sample is stable up to 24 h at room temperature and for several days at 4°C. Dilute serum (1:20) and urine samples (1: 1000) before use.

16.3 Principle

Urea reacts directly with diacetyl monoxime under strong acidic conditions to give a yellow-colored condensation product. Firstly, diacetyl is released from diacetyl monoxime, and then urea reacts with diacetyl in hot acidic medium to give diazine complex. Diazine is stabilized as pink-colored compound by thiosemicarbazide and ferric ions. The acid reagent of H_3PO_4 and H_2SO_4 acid is used to provide acidic medium for these reactions, and final pink product has absorption maximum at 520 nm in proportion to amount of urea.

16.4 Reagents

1. **Acid reagent**
   Sol. A – Dissolve 0.5 g of ferric chloride ($FeCl_3$) in 2 ml of distilled water. Add 10 ml H_3PO_4 acid to it. Then mix and make final volume to 25 ml with distilled water. Store at room temperature in a brown bottle.
   Sol. B – Prepare 20% aqueous solution of H_2SO_4. Then mix 0.25 ml of solution A with 500 ml of solution B to prepare acid reagent.
2. **Color reagent**
   Sol. A – Prepare 2% diacetyl monoxime solution in distilled water.
   Sol. B – Prepare 0.5% thiosemicarbazide solution in distilled water.

Mix 35 ml of solution A with 35 ml of solution B, and make up to 500 ml with distilled water. Store in a brown bottle at room temperature.

3. **Stock urea standard**: Prepare 100 mg/dl urea solution in distilled water.
4. **Working standard**: Prepare 2.5 mg/dl working solution.

16.5 Procedure

1. Take a set of nine test tubes and label as blank, standard (S_1–S_6), test for serum sample (Ts), and test for urine sample (Tu).
2. Add 0.1, 0.2, 0.4, 0.6, 0.8, and 1.0 ml of working urea standard into the tubes labeled as S_1 to S_6, respectively, and make the final volume 2 ml with distilled water.
3. In Ts add 0.2 ml diluted serum sample (1:10) and 0.4 ml of diluted urine sample (1:100) to the tube marked as T_U, and make the volume of each tube to 2 ml with distilled water.
4. Add 2 ml of distilled water in blank tube, and then add 2 ml each of acid reagent and color reagent to all tubes.
5. Mix well and incubate all the tubes in boiling water bath for 20°C and cool.
6. Measure the absorbance at 520 nm in a spectrophotometer.

Reagents	Blank	S_1	S_2	S_3	S_4	S_5	S_6	T_S	T_U
Working urea (ml)	–	0.1	0.2	0.4	0.6	0.8	1.0	–	–
Amount of urea (mg)	–	0.0025	0.005	0.010	0.015	0.020	0.025		
Serum/urine (ml)	–	–	–	–	–	–	–	0.2	0.4
Distilled water (ml)	2	1.9	1.8	1.6	1.4	1.2	1	1.8	1.6
Acid reagent (ml)	2	2	2	2	2	2	2	2	2
Color reagent (ml)	2	2	2	2	2	2	2	2	2

Mix well and keep in boiling water bath for 20 min, and cool and read absorbance at 520 nm

16.6 Calculations

Plot a graph between amount of urea at x-axis and absorbance at y-axis. Then extrapolate amount of urea from the given serum/urine sample from the graph (include dilution factor for calculation). The urea concentration can also be calculated by using the following equation:

Amount of urea in blood
$$= \frac{\text{OD of test} \times \text{amount of standard (mg)} \times 100 \times \text{dilution factor}}{\text{OD of standard} \times \text{volume of sample (ml)} \times 1000} = x\,g/dl$$

Amount of urea in urine
$$= \frac{\text{OD of test} \times \text{amount of standard (mg)} \times 1000 \times \text{dilution factor}}{\text{OD of standard} \times \text{volume of sample (ml)} \times 1000} = x\,g/L$$

16.7 Clinical Significance

The normal range of blood urea is 15–40 mg/dl, and urinary excretion is 15–40 g/day. Increase in blood urea is called uremia, and it occurs in wide variety of diseases. Uremia is of three types:

Prerenal Uremia It is caused by mild dehydration, high protein catabolism, muscle wasting as in starvation, reabsorption of blood protein after a gastrointestinal hemorrhage, treatment with cortisol, and decreased perfusion of kidneys.

Renal Uremia The blood urea is increased in all forms of kidney diseases. In acute glomerulonephritis values goes up to 300%.

Postrenal Uremia It is caused by conditions such as nephrolithiasis, enlargement of prostate gland, tumors of the urinary bladder, and stones in ureters. These conditions obstruct the urine outflow through ureters, bladder, or urethra.

Decrease in blood urea is rare. It has been reported in some cases of severe diseases, e.g., viral hepatitis with extensive necrosis. Levels also decrease in low protein diet intake. In pregnancy, blood urea level is lower (15–20 mg/dl).

To Determine Urea Clearance 17

17.1 Theory

To assess the kidney functions, clearance of some of the substances from the kidneys is measured. The clearance of substances that are filtered exclusively or predominantly by the glomeruli but neither reabsorbed nor secreted by other regions of the nephron can be used to measure glomerular filtration rate (GFR). The concept of urea clearance was introduced by Van Slyke and his coworkers in 1928. Urea clearance (C) is defined as volume of blood or plasma completely cleared of urea/min by the kidneys and can be expressed mathematically as

$$C = UV/P$$

U = concentration of urea in urine/dl
P = concentration of urea in plasma/dl
V = volume of urine excreted/min

17.2 Maximum Urea Clearance

If the volume of urine excreted/min is more than 2 ml/min, maximum urea clearance is calculated, and the observed maximum clearance can be compared with the average normal value of 75 ml/min.

$$\text{Maximum urea clearance } (\%\text{average normal}) = \frac{\text{observed clearance} \times 100}{75}$$
$$= 1.33 \times UV/P$$

17.3 Standard Urea Clearance

If the volume of urine excreted/min is less than 2 ml/min, standard urea clearance is calculated by comparing the observed clearance after adjustment for "U" with a standard of 54 ml/min, the average normal urea clearance.

$$\text{Standard urea clearance (\% average normal)} = \frac{\text{observed clearance} \times 100}{54}$$

For any value of "U," the urea clearance is finally expressed as % of average normal value.

17.4 Procedure

Urea clearance can be carried either from 24 h urine specimen or timed urine specimen. 24 h urine specimen is collected in a bottle containing a preservative (10 ml of 6 N HCl). In between the 24 h period, collect a blood sample also for estimation of urea in plasma. After the collection of 24 h urine specimen, measure the volume of urine to calculate volume of urine excreted/min.

If timed urine specimen is to be used, give a cupful (200–400 ml) water to drink, and ask the patient to empty the bladder and note the time. Exactly after 1 h, collect the first urine specimen, and mark it as urine sample "1". After 5 min of collection of urine sample, collect a blood sample for urea estimation. Exactly after 2 h of intake of water, ask the patient to empty the bladder completely, and mark that specimen as urine sample "2". Measure the volume of both the urine samples, and estimate the concentration of urea in plasma and both the urine samples. Then calculate standard or maximum urea clearance from both the urine samples according to the volume of urine excreted/min.

17.5 Precautions

1. Collect exactly 24 h urine sample, and note the volume accurately.
2. Use urine preservative for the collection of 24 h urine specimen.
3. If timed urine specimen is collected, we can check the reliability of urine collection. The differences between the 1st and 2nd hourly specimens should not exceed 10%.

17.6 Clinical Significance

There is wide range of urea clearance in normal persons. Normal maximum clearance varies from 60 to 95 ml/min, with an average of 75 ml/min. The normal standard urea clearance varies from 40 to 65 ml/min with an average of 54 ml/min.

17.6 Clinical Significance

Low-protein diet may give low clearance values. In prerenal uremia, there is fall in urea clearance. Increase in renal failure decreases clearance value. The clearance may fall to 50% of normal, while plasma urea will be within normal range. With further reduction in clearance, plasma urea levels rise. With more severe renal failure, the clearance may be less than 20% of normal.

To Estimate Creatinine Level in Serum and Urine by Jaffe's Reaction

18.1 Theory

Creatinine is a waste product derived from dehydration of creatine and is excreted by the kidneys. It is synthesized in the kidneys, liver, and pancreas from three amino acids –arginine, glycine, and methionine – by two enzyme-mediated reactions. After synthesis, creatine is transported in the blood to other organs such as the muscle and brain and phosphorylated to phosphocreatine which is a high-energy compound. Some of the free creatine in the muscle is spontaneously converted to creatinine (anhydride of creatine). Creatine and creatine phosphate constitute about 400 mg/100 mg of fresh muscle. Both compounds are converted spontaneously to creatinine at rate of 1–2% (Fig. 18.1).

18.2 Specimen Requirements

Serum or plasma (fluoride or heparin) is stable for 24 h if refrigerated.

Urine: Collect 24 h urine sample. The urine should be kept in cool place (stable up to 4 days if refrigerated). Dilute urine sample 1:10 before use.

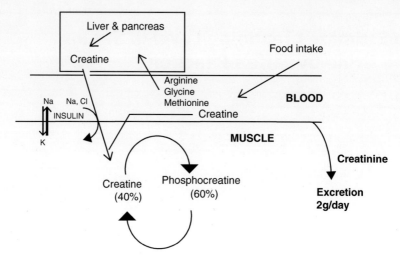

Fig. 18.1 Creatinine metabolism

18.3 Principle

Creatinine present in serum directly reacts with alkaline picrate resulting in the formation of a red-colored tautomer of creatinine picrate, the intensity of which is measured at 520 nm. The color produced by creatinine fades after acidifying with 0.25 ml of 0.1 N H_2SO_4 but remains if produced by non-specific chromogens.

18.4 Reagents

1. **2/3 N H_2SO_4**
2. **10% sodium tungstate:** Dissolve 10 g sodium tungstate in final volume of 100 ml distilled water.
3. **0.04 M picric acid:** Dissolve 9.3 g pure dried picric acid in water, and make final volume to 1000 ml.
4. **0.75 N NaOH**
5. **Stock creatinine standard 100 mg/dl:** Dissolve 100 mg of pure creatinine in 100 ml of 0.1 N HCl.
6. **Working creatinine standard:** Prepare 1 mg/dl solution in distilled water.

18.5 Preparation of Protein-Free Filtrate

For serum add 4 ml of 2/3 N H_2SO_4 to 0.5 ml of serum, and add 0.5 ml of 10% sodium tungstate solution. Mix well and keep for 10 min. Centrifuge at 3000 rpm for 5 min. Supernatant is used for the estimation of creatinine.

For urine: dilute urine 1:10 with distilled water. Prepare filtrate as with serum.

18.6 Procedure

1. Take a set of nine test tubes, and label as blank, standard (S_1–S_6), test for serum (Ts), and test for urine sample (Tu).
2. Add 0.5 ml, 1 ml, 1.5 ml, 2 ml, 2.5 ml, and 3 ml of working 1 mg/dl creatinine standard to the tubes labelled as S_1 to S_6, respectively, and make the volume 3 ml with distilled water. To blank tube, add 3 ml of distilled water.
3. Add 3 ml of protein-free filtrate of serum and urine samples in respective tubes followed by 1 ml each of NaOH and picric acid.

Reagents	Blank	S_1	S_2	S_3	S_4	S_5	S_6	T_S	T_U
Creatinine working standard (ml)	–	0.5	1	1.5	2	2.5	3	–	–
Creatinine amount (µg)	–	5	10	15	20	25	30	–	–
Protein-free filtrate of serum/urine (ml)	–	–	–	–	–	–	–	3	3
Distilled water (ml)	3	2.5	2	1.5	1	0.5	–	–	–
0.75 N NaOH (ml)	1	1	1	1	1	1	1	1	1
Picric acid (ml)	1	1	1	1	1	1	1	1	1

Mix well and keep for 10 min at room temperature; read absorbance at 520 nm

18.7 Calculations

Plot a graph between amount of creatinine at x-axis and absorbance at y-axis. Extrapolate creatinine concentration of given serum/urine sample from the graph (include dilution factor for calculation). The creatinine concentration can also be calculated by using equation:

$$\text{Amount of creatinine in blood} = \frac{\text{OD of test} \times \text{amount of standard}\,(\mu g) \times 100}{\text{OD of standard} \times \text{volume of sample}\,(\text{ml}) \times 1000}$$
$$= x\,\text{mg/dl}$$

Amount of creatinine in urine
$$= \frac{\text{OD of test} \times \text{amount of standard}\,(\mu g) \times 1000 \times \text{dilution factor}}{\text{OD of standard} \times \text{volume of sample}\,(\text{ml}) \times 1000 \times 1000} = x\,\text{g/L}$$

Note 3 ml of protein-free filtrate is equivalent to 0.3 ml serum/urine sample since 0.5 ml sample was made to 5 ml before preparing protein-free filtrate.

18.8 Clinical Significance

Normal range of creatinine in serum is 0.6–1.5 mg/dl and in urine is 90–150 mg/l of urine. It increases in renal disease, and creatinine value goes up to 20 mg/dl in muscular dystrophy and different types of muscular tissue diseases. Serum creatinine level also increases in fever and starvation. There is considerable muscle wasting in diabetes mellitus resulting in high creatinine levels. Creatinine levels are also raised in urinary tract obstruction, due to blockage of creatinine excretion through urine. Decrease in serum creatinine levels is seen in decreased muscle mass with aging and in pregnancy.

To Determine Creatinine Clearance

19.1 Theory

Creatinine clearance is defined as the volume of plasma from which creatinine is completely cleared/min by the kidneys. Creatinine clearance (C) is expressed mathematically as

$$C = UV/P$$

U = Concentration of creatinine in urine/dl
P = Concentration of creatinine in plasma/dl
V = Volume of urine exerted (ml/min)

Creatinine formed during muscular metabolism is eliminated from plasma mainly by glomerular filtration; thus measurement of creatinine clearance is an approach to find GFR.

19.2 Specimen Requirement and Procedure

In order to estimate creatinine clearance, a 24 h urine sample is collected. The patient is asked to void urine at a specified time, which is then discarded. Subsequent urine samples for the next 24 h are collected during precisely timed intervals (at 4, 12, or 24 h). The final volume is measured and used to calculate value of "V." Keep samples refrigerated or on ice during collection and refrigerate until analysis. Creatinine concentration is then estimated in urine and serum by Jaffe reaction.

19.3 Clinical Significance

Normal range of creatinine clearance for male is 95–125 ml/min and for female is 85–125 ml/min. Creatinine clearance is a good indicator of GFR as compared to urea clearance because creatinine is not reabsorbed from the kidney tubules but urea is reabsorbed from the kidney tubules. The creatinine clearance value is decreased in case of renal failure/glomerular damage (serum level might be normal) and increased in case of muscular disorders. Proper hydration of the patient, ensuring a urine flow of 2 ml/min or more results, is a more precise measure of the GFR. If an average clearance value of 125 ml/min is taken as 100% clearance, then renal damage may be assessed as follows:

Up to 70% clearance	- Normal
70–50%	- Mild renal damage
49–20%	- Moderate damage
Below 20%	- Severe damage

To Determine the Uric Acid Concentration in Serum and Urine

20.1 Theory

Uric acid is the end product of purine metabolism (adenine and guanine) present in RNA and DNA. Uric acid is 2,6,8-trihydroxypurine and exists in keto and enol forms. It exists only as the monosodium and disodium salt. In plasma, uric acid is present as monosodium salt. The pool of uric acid in the body is about 1200 mg, out of which about half amount undergoes turnover daily, that is, about 600 mg/dl is formed daily and about the same amount is lost, of which about 75% is exerted through urine and about 25% destroyed by bacteria in the colon.

Structure of uric acid

20.2 Specimen Requirements

Serum is stable for 3–5 days at 4°C and for 6 months at −20°C. Do not use oxalate for phosphotungstate method. For uricase method, do not use EDTA or fluoride. For urine sample, collect 24 h sample. Do not refrigerate and add NaOH to keep urine alkaline.

20.3 Principle

Proteins are precipitated with tungstic acid and are removed by centrifugation. Protein-free filtrate of serum or urine containing uric acid is oxidized to allantoin by phosphotungstic acid and is reduced to tungsten blue in alkaline medium. The blue color thus produced is measured at 700 nm and compared with that of uric acid standard.

$$\text{Uric acid} + \text{Phosphotungstic acid} + O_2 + 2H_2O \rightarrow \text{Allantoin} + CO_2$$
$$\text{Allantoin} \xrightarrow[\text{reduction}]{\text{alkali}} \text{Tungsten blue}$$

20.4 Reagents

1. **10% sodium tungstate aqueous solution.**
2. **3/2 N H_2SO_4**
3. **Phosphotungstic acid**: Dissolve 10 g of sodium tungstate in about 150 ml of distilled water. Add 8 ml of 85% H_3PO_4 and put some glass beads. Reflux for 2–3 h. Cool to room temperature and then add 50 ml of distilled water containing 8 g of dissolved lithium sulfate. Mix and make final volume up to 250 ml with distilled water and store in colored bottle.
4. **7% aqueous solution of Na_2CO_3.**
5. **Uric acid standard (20 mg/dl)**: Dissolve 12 mg of lithium carbonate in warm water and add 20 mg of uric acid and dissolve by heating. Make volume up to 100 ml with distilled water.

20.5 Procedure

Protein-free filtrate of serum is prepared by adding 0.5 ml serum in a test tube containing 4 ml of 3/2 N H_2SO_4 and added with 0.5 ml of 10% sodium tungstate. Mix well and keep for 5 min and centrifuge at 3000 rpm for 5 min.

Protein-free filtrate of urine is prepared by diluting urine 1:10 with distilled water and following same process as that of serum. Add uric acid standard in tubes labelled as (S_1–S_6) and mix other reagents according to protocol given in table.

Reagents	Blank	S1	S2	S3	S4	S5	S6	TS	TU
Working uric acid standard (μl)	–	50	100	150	200	250	500	–	–
Uric acid amount (μg)	--	10	20	30	40	50	100	–	–
Protein-free filtrate of serum/urine (ml)	–	–	–	–	–	–	–	3	3
Distilled water (ml)	3	2.95	2.90	2.85	2.80	2.75	2.5	–	–

(continued)

Na$_2$CO$_3$ (ml)	1	1	1	1	1	1	1	1	1
Phosphotungstic acid (ml)	1	1	1	1	1	1	1	1	1

Mix well and keep for 20 min at room temperature, read absorbance at 700 nm

20.6 Calculations

Plot a graph between amount of uric acid at x-axis and absorbance at y-axis and extrapolate uric acid concentration of given sample from the graph. The uric acid concentration can also be calculated by using equation:

$$\text{Uric acid amount in urine} = \frac{\text{OD of test} \times \text{amount of standard} (\mu g) \times 100}{\text{OD of standard} \times \text{volume of sample} \times 1000}$$
$$= x \text{ mg/dl}$$

Uric acid amount in urine
$$= \frac{\text{OD of test} \times \text{amount of standard} (\mu g) \times \text{dilution factor} \times 1000}{\text{OD of standard} \times \text{volume of sample} \times 1000} = x \text{ mg/L}$$

Note 3 ml of protein-free filtrate is equivalent to 0.3 ml serum/urine sample since 0.5 ml sample was made to total 5 ml before preparing protein-free filtrate.

20.7 Clinical Significance

Uric acid levels in serum range from 3.5–8 mg/dl in males and 2.0–6.5 mg/dl in females. Normal uric acid excretion through urine depends upon diet. A purine-rich diet and severe exercise increase uric acid levels. Estimation of uric acid in serum is helpful in the diagnosis of several pathological conditions particularly gout where increased uric acid levels up to 6.5–12 mg/dl is observed. Gout is the most common disease that increases uric acid levels in serum. It is disease of joints in which uric acid is deposited as crystals. Increased uric acid in blood is generally accompanied by increased excretion in urine. Other disease conditions such as renal failure, leukemia, multiple myeloma, lymphoma, glycogenesis, chronic hemolytic anemia, and pernicious anemia lymphosarcoma also increase serum uric acid levels. Serum uric acid levels less than 2 mg/dl cause hypouricemia. It is caused due to severe hepatocellular disease and defective renal tubular reabsorption of uric acid. Low values may be also found in Wilson's disease. Several drugs affect the handling of urate by the body. Urinary decrease in uric acid levels is observed in xanthinuria, folic acid deficiency, or lead toxicity.

20.8 Precautions

1. Molybdate-free sodium tungstate is used for the preparation of phosphotungstic acid.
2. Store uric acid standard in refrigerator.

Estimation of Total Calcium in Serum and Urine

21.1 Theory

Calcium is the most abundant mineral cation in the body which contains 1–1.5 kg of total body weight in the adults. Over 99% is present in the bones and teeth. A small fraction is also present outside the skeletal tissue and performs wide variety of functions. Calcium is vitally important in maintaining the correct conditions for normal neuromuscular transmission and glandular secretions and for the activity of enzyme systems. Most of blood calcium is present in plasma. There is almost no calcium in red blood cells and other intracellular fluid. About 50% of total plasma calcium exists in ionized form which is functionally most active. About 40% of plasma calcium is protein bound, and 10% is complexed with citrate, organic acid, or bicarbonates. About 90% of protein-bound calcium is associated with albumin, and the remaining 10% is associated with globulins. Ionized and citrate-bound calcium is diffusible from blood to the tissues, while protein bound is nondiffusible (Fig. 21.1). In the laboratory, all three fractions of calcium are measured together. Calcium binds to negatively charged sites on proteins which is pH dependent. Alkaline conditions promote calcium binding and decrease free calcium, whereas acidic conditions decrease calcium binding and increase free calcium levels.

21.2 Specimen Requirements

Serum/plasma (fasting) sample is required. Take blood with minimal venous occlusion and without exercise. For urine sample, collect 24 h urine in the container containing 10 ml HCl (6 mol/L) or acidify after collection to pH < 2.0 to dissolve calcium salts.

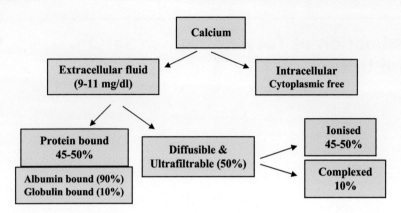

Fig. 21.1 Distribution of calcium in body fluids

21.3 Method

Calcium is estimated by O-Cresolphthalien Complexone method.

21.4 Principle

Hydrochloric acid dissociates calcium from protein. Then total calcium reacts with ortho-cresolphthalein complex to form a purple-colored complex in alkaline medium. Diethylamine provides the alkaline pH, and 8-hydroxy quinoline prevents the interference of magnesium ions. The color thus produced is compared with that of calcium standard.

21.5 Reagents

1. **Cresolphthalein reagent**: Dissolve 70 mg of ortho-cresolphthalein complexone in 16.5 ml conc. HCl. Then add 2.5 g 8-hydroxy quinoline, mix and dilute to 1000 ml with distilled water.
2. **Diethylamine buffer (3.75% v/v) in distilled water**: Mix 3.75 ml diethylamine in a final volume of 100 ml distilled water.
3. **Calcium Standard 4 mg %**: Dissolve 10 mg of $CaCO_3$ in 0.1 N HCl and make volume 100 ml with 0.1 N HCl solution.

21.6 Procedure

1. Take a set of 10 test tubes in duplicate and mark as blank, standard (S_1–S_8), test for serum (Ts), and test for urine (T_U) sample.

2. Pipette out 25, 50, 75, 100, 125, 150, 175, and 200 µl of 4 mg % calcium standard into the tubes labelled as S_1 to S_6. Add 50 µl undiluted serum and urine samples into tubes Ts and T_U.
3. Then make the final volume of all the tubes 200 µl by adding distilled water. Add 2 ml cresolphthalein complexone reagent to all the tubes and mix well.
4. Then add 2.5 ml diethylamine buffer to all the tubes, mix well, and measure absorbance at 575 nm.

Reagents	Blank	S_1	S_2	S_3	S_4	S_5	S_6	S_7	S_8	T_S	T_U
Calcium standard (µl)	–	25	50	75	100	125	150	175	200	–	–
Calcium amount (µg)	–	1	2	3	4	5	6	7	8	–	–
Serum/urine (µl)	–	–	–	–	–	–	–	–	–	50	50
Distilled water (µl)	200	175	150	125	100	75	50	25	–	150	150
Cresolphthalein complexone reagent (ml)	2	2	2	2	2	2	2	2	2	2	2
Diethylamine buffer (ml)	2.5	2.5	2.5	2.5	2.5	2.5	2.5	2.5	2.5	2.5	2.5

Mix well and measure absorbance at 575 nm

21.7 Calculations

Plot a graph by taking calcium amount at x-axis vs absorbance at y-axis, and extrapolate calcium concentration of given serum/urine sample from the graph. The calcium levels in serum and urine can also be calculated by using equation:

$$\text{In serum} = \frac{\text{OD of test} \times \text{amount of standard} (\mu g) \times 100}{\text{OD of standard} \times \text{volume of sample} \times 1000}$$
$$= x \, mg/dl$$

$$\text{In urine} = \frac{\text{OD of test} \times \text{amount of standard} (\mu g) \times 1000}{\text{OD of standard} \times \text{volume of sample} \times 1000}$$
$$= x \, mg/L$$

21.8 Clinical Significance

The normal value of calcium in serum is 9–11 mg/dl and 100–300 mg/24 h urine. Extracellular calcium provides calcium ions for maintenance of intracellular calcium, bone mineralization, blood coagulation, and maintenance of plasma membrane potential. Calcium levels are increased in hyperparathyroidism (determination of ionized serum calcium is more useful for the diagnosis of hyperparathyroidism), parathyroid hormone injection, hypervitaminosis "D," prolonged intake of milk and alkali (the milk-alkali syndrome), multiple myeloma, and polycythemia. It is also increased in acute and chronic renal failure and osteomalacia with malabsorption. Hypocalcemia occurs when serum calcium levels fall below 7 mg/dl. Hypocalcemia is more serious and life-threatening condition than hypercalcemia. Calcium levels are decreased in hypoparathyroidism, osteomalacia, hyperphosphatemia, tetany, renal failure, and nephrotic syndrome. In rickets, the product of serum calcium and phosphorus decreases usually below 30 mg/dl. An increase in alkaline phosphatase activity is a characteristic feature of rickets. It is necessary to measure total serum proteins and albumin levels simultaneously for proper interpretation of serum calcium levels, since 0.8 mg of calcium is bound to 1.0 g of albumin in serum. To correct, add 0.8 mg/dl for every 1 g/dl that serum albumin falls below 4 g/dl.

Estimation of Inorganic Phosphorus in Serum and Urine

22.1 Theory

The body contains about 800 g of phosphorus. Of this, 90% is present in bones, so it is more widely distributed in the body than calcium. In the soft tissues, phosphorus is present as organic form and incorporated into macromolecules. Phosphorous is also present in some proteins, lipids, and nucleic acid and in some coenzymes. Phosphorus also plays a role in acid-base regulation of body fluids particularly by the kidneys. The phosphorus of the blood is of four types:

1. Inorganic phosphorus – the phosphates of alkaline and alkali earth metals, present as H_2PO_4 and HPO_4^{2-}
2. Organic or ester phosphorus – including glycerophosphates, nucleotide phosphate, hexose phosphate, etc.
3. Lipid phosphorus – present in lecithin, cephalin, sphingomyelin
4. Residual phosphorus present in small amounts

Out of four types explained above, it is the inorganic phosphorus present in serum which is measured. Phosphate in serum exists as both the monovalent and divalent phosphate anions. The ratio of $H_2PO_4^-$ and HPO_4^{2-} is pH dependent. About 10% of phosphate in serum is bound to proteins; 35% is complexed with sodium, calcium, and magnesium; and the remainder is free. Inorganic phosphate also forms major component of hydroxyapatite in the bone, thereby playing an important part in structural support of the body and providing phosphate for the extracellular and intracellular pool.

22.2 Specimen Requirements

Fasting serum sample is preferred. Avoid hemolysis since blood cell contains high amount of phosphate (RBCs contain about 40 mg/dl of phosphorus). Sample is stable for several days at 4°C. For urine collect 24 h samples in acid-washed detergent-free container, and acidify with HCl after collection. Stable for 6 months.

22.3 Methodology

Phosphorus content is measured by following the method of Fiske and Subbarow.

22.4 Principle

Proteins are precipitated with TCA and are removed by centrifugation. Protein-free filtrate contains phosphorus that reacts with molybdic acid to form phosphomolybdic acid. The hexavalent molybdenum of the phosphomolybdic acid is reduced by 1,2,4 amino-naphthol-sulfonic acid (ANSA). The color thus produced is directly proportional to the amount of phosphorous present in the sample.

22.5 Reagents

1. **20% TCA**: Dissolve 20 g of TCA in 100 ml distilled water.
2. **Ammonium molybdate reagent**
   Sol. A: 1 N HCl–11.5 ml conc. HCl is diluted to 100 ml with distilled water.
   Sol. B: Ammonium molybdate solution – Dissolve 6.25 g of ammonium molybdate in water, and make volume to 150 ml with distilled water.
   Prepare ammonium molybdate reagent by mixing Sol. A and B with constant mixing.
3. **ANSA reagent**: Dissolve 250 mg of ANSA powder in a solution containing 97.5 ml of 15% aqueous solution of sodium bisulfite and 2.5 ml of 20% sodium sulfite.
4. **Phosphorus standard (20 mg %)**: Dissolve 11.48 mg of $Na_2HPO_4.2H_2O$ in distilled water to make final volume 100ml.

22.6 Preparation of Protein-Free Filtrates

For Serum Mix 0.5 ml serum with 4.5 ml 20% TCA and allow standing for 10 min at room temperature. Then centrifuge at 3000 rpm for 5 min and collect supernatant.

For Urine Dilute urine 1:10 with distilled water and prepare protein-free filtrate the same way as for serum.

22.7 Procedure

1. Take a set of nine test tubes, and label them as blank, standard (S_1–S_6), test for serum sample (T_S), and test for urine sample (T_U).
2. Add 10, 20, 40, 60, 80, and 100 μl of 20 mg % phosphorus standard in tubes labelled as S_1 to S_6, respectively, and make the final volume 2.5 ml with 20% TCA. In blank add 2.5 ml of 20% TCA.
3. Add 2.5 ml of protein-free filtrate of serum and urine into the tubes labelled at T_S and T_U, respectively.
4. Add 1 ml of ammonium molybdate reagent and 0.2 ml ANSA reagent to all the tubes and vortex. Keep at room temperature for 20 min and read absorbance at 700 nm.

Reagents	Blank	S_1	S_2	S_3	S_4	S_5	S_6	T_S	T_U
Phosphorus standard, 20 mg % (μl)	–	10	20	40	60	80	100	–	–
Phosphorus amount (μg)	–	2	4	8	12	16	20	–	–
Protein-free filtrate of serum/urine (ml)	–	–	–	–	–	–	–	2.5	2.5
20% TCA (ml)	2.5	2.49	2.48	2.46	2.44	2.42	2.40	–	–
Ammonium molybdate (ml)	1	1	1	1	1	1	1	1	1
ANSA (ml)	0.2	0.2	0.2	0.2	0.2	0.2	0.2	0.2	0.2

Mix well and keep for 20 min at room temperature, read absorbance at 700 nm

22.8 Calculations

Plot a graph by taking phosphorus amount at x-axis vs absorbance at y-axis, and extrapolate its concentration from the given serum/urine sample from the graph. **The amount of phosphorus extrapolated from graph will be in 250 μl of serum/urine sample since 2.5 ml of protein-free filtrate was used for analysis.** The phosphorus amount can also be calculated in serum and urine samples by using equation:

$$\text{Phosphorus amount in serum} = \frac{\text{OD of test} \times \text{amount of standard}\,(\mu g) \times 100}{\text{OD of standard} \times \text{volume of sample}\,(0.25\,\text{ml}) \times 1000} = x\,\text{mg/dl}$$

Phosphorus amount in urine

$$= \frac{\text{OD of test} \times \text{amount of standard } (\mu g) \times \text{dilution factor} \times 1000}{\text{OD of standard} \times \text{volume of sample} \times 1000^* \times 1000^*} = x \text{ g/L}$$

*To convert μg to gram

22.9 Clinical Significance

Normal value of serum inorganic phosphorus is 3.0–4.5 mg/dl in adults and is 3–5.5 mg/dl in children. Normal range for urine sample is 0.5–1.5 g/24 h. Increase in an inorganic phosphorus amount is observed in acute and chronic renal failure, hypoparathyroidism, cell lysis, respiratory acidosis, untreated diabetes, ketoacidosis, increased phosphate intake, use of phosphate containing laxative and decreased renal phosphate exertion due to decreased GFR, and increased tubular reabsorption or acromegaly. The decrease in inorganic phosphorus levels occurs in rickets (in osteomalacia the phosphorus levels are lower than rickets), lowered renal phosphate threshold, decrease in intestinal phosphate absorption due to increased loss in vomiting, diarrhea, the use of phosphate-binding antacid, vitamin D deficiency, and in intracellular phosphate loss due to acidosis – ketoacidosis and lactoacidosis.

22.10 Precautions

1. All the glasswares used for the estimation and preparation of reagents should be dipped in HNO_3 (5 N) for 24 h and washed thoroughly in distilled water.
2. Urine should be preserved in acidic conditions by adding 10 ml HCl to 24 h urine collection. In alkaline condition the phosphates tends to precipitate.
3. Collect blood without hemolysis and separate serum immediately after it is formed.

To Estimate the Amount of Total Cholesterol in Serum

23.1 Theory

Cholesterol is a solid alcohol found exclusively in animals and contains 27 carbon atoms ($C_{27}H_{46}O$). It possesses a ring structure called cyclopentanoperhydrophenanthrene. It is a major component of biological membranes and lipoproteins and amphipathic in nature. It is a precursor in many metabolic pathways including synthesis of vitamin D, steroid hormones, and bile acids. About 400–700 mg of cholesterol present in the gut is contributed by biliary secretion and the turnover of mucosal cells. The cholesterol present in the intestine is the unesterified form. This form is solubilized by the formation of mixed micelles that contain the unesterified cholesterol, fatty acids, monoglycerides, phospholipids, and conjugated bile acids. Maximum absorption of cholesterol occurs in small intestine. Once absorbed into the mucosal cells, cholesterol along with triglycerides, phospholipids, and a number of specific apoproteins is reassembled into a large micelle called chylomicrons. Chylomicrons enter the lymphatics, which empty into the thoracic duct and eventually enter the systemic venous circulation. Total cholesterol contains free cholesterol and esterified cholesterol in the ratio 1:3.

Cholesterol

23.2 Specimen Requirements

Serum (fasting) or plasma (EDTA or heparin may be used but not oxalate, fluoride, or citrate). Serum is stable for 5–7 days at 4 °C. Results with EDTA plasma are 3% lower than serum results.

23.3 Methodology

Cholesterol is estimated by Zak ferric chloride- sulfuric acid method.

23.4 Principle

Ferric chloride in acetic acid precipitates the serum proteins, and the cholesterol liberated reacts with conc. H_2SO_4 to make a chromophore (colored complex). The reaction seems to involve oxidation, dehydration, and sulfonation. The amount of color produced is directly proportional to the cholesterol present in the sample which is measured at 550 nm. The reaction involves "3-hydroxy-5ene" part of the cholesterol molecule, which is first dehydrated to form cholesta 3,5-diene and then oxidized by the H_2SO_4 to link two molecules together as bis-cholesta-3,5-diene. This compound can be sulfonated by the H_2SO_4 to produce mono- and disulfonic acids which are highly colored. In presence of more H_2SO_4 and ferric ions, the disulfonic acids are preferentially formed and are red in color. The color thus produced is measured at 550 nm.

23.5 Reagents

1. **Precipitating reagent**: 0.05% $FeCl_3$ in glacial acetic acid
2. **Conc. H_2SO_4**
3. **Cholesterol standard**: 100 mg % in glacial acetic acid

23.6 Procedure

1. Take a set of seven test tubes and label them as blank, standard (S_1-S_5), and test (T).
2. Dispense 20, 40, 60, 80, and 100 µl of 100 mg % cholesterol standard into the tubes labelled as (S_1-S_5), respectively, and make the volume 3 ml with precipitating reagent.
3. In the blank tube, add 3 ml precipitating reagent.

4. Add 0.05 ml serum in 5 ml of precipitating reagent, mix well and keep for 5 min, and centrifuge it at 3000 rpm for 5 min. Take 3 ml of supernatant and add to the tube labelled as T.
5. To each tube add 2 ml of conc. H_2SO_4 through the sides of the tube. Mix well immediately. Incubate for 20 min at room temperature and read absorbance at 550 nm.

Reagents	Blank	S_1	S_2	S_3	S_4	S_5	T
Cholesterol standard (µl)	–	20	40	60	80	100	–
Cholesterol amount (µg)	–	20	40	60	80	100	–
Precipitating reagent (ml)	3	2.98	2.96	2.94	2.92	2.90	–
Protein-free filtrate (ml)	–	–	–	–	–	–	3
Conc. H_2SO_4 (ml)	2	2	2	2	2	2	2
Overlay and mixed thoroughly and keep at room temperature for 20 min							

23.7 Calculations

Plot a graph by taking cholesterol amount at x-axis vs absorbance at y-axis, and extrapolate cholesterol concentration of the given serum sample from the graph. **The amount of cholesterol extrapolated from graph will be in 0.03 ml of serum sample since 0.05 ml of serum sample was mixed with precipitating reagent and 3 ml of this was used for cholesterol analysis.** The cholesterol amount in serum sample can also be calculated by using equation:

$$\text{Cholesterol in serum} = \frac{\text{OD of test} \times \text{amount of standard} (\mu g) \times 100}{\text{OD of standard} \times \text{volume of sample} (0.03 \text{ ml}) \times 1000}$$
$$= x \, \text{mg/dl}$$

23.8 Clinical Significance

Normal range of total cholesterol in blood ranges from 150 to 240 mg/dl. Serum cholesterol rises with age. Hypercholesterolemia occurs in nephrotic syndrome, diabetes mellitus, obstructive jaundice (due to obstruction of large ducts), coronary thrombosis, angina pectoris, atherosclerosis, and other heart disease. In pregnancy, the increase in cholesterol levels may reach 20–25% of the normal value. Hypercholesterolemia occurs in hypothyroidism, liver damage, pernicious anemia, hemolytic jaundice, abetalipoproteinemia, and cortisone administration.

To Estimate Total and Direct Bilirubin in Serum

24.1 Theory

Bilirubin originates from breakdown of heme. It is a waste product and the body eliminates it through bile. The heme present in erythrocytes contributes approximately 85% of total bilirubin which is destroyed in the reticuloendothelial cells. The destruction of red blood cell precursors in the bone marrow and catabolism of other heme-containing proteins like myoglobin, cytochromes, and peroxidases release the remaining 15% of bilirubin. Biliverdin is first formed from the porphyrin part of heme in reticuloendothelial cells and is reduced to bilirubin called unconjugated bilirubin, which is then transported to the liver in association with albumin. This unconjugated bilirubin is insoluble in water. Bilirubin then enters to microsomes of hepatocytes and conjugated by the action of glucuronyl transferase to produce bilirubin monoglucuronide and diglucuronide (conjugated bilirubin). The bilirubin conjugated with glucuronic acid is water soluble and it exerted through bile. The conjugated bilirubin present in bile passes along the bile ducts into the intestine. Here it is reduced by bacterial action and is also deconjugated, mainly in the colon to "urobilinogens." Urobilinogen is recycled through the body and a part of it is excreted through the urine (Fig. 24.1).

24.2 Specimen Requirements

Serum is a desirable sample. Protect the sample from light since exposure to white or UV light may decrease total and indirect bilirubin up to 20%. Hemolysis increases indirect bilirubin.

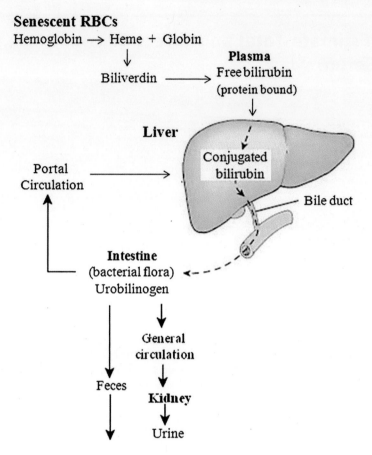

Fig. 24.1 Bilirubin metabolism and elimination

24.3 Principle

Bilirubin reacts with diazotized sulfanilic acid to produce a red-purple-colored compound called azobilirubin under acidic conditions. The color thus produced is measured in a colorimeter at 540 nm and compared with that of bilirubin standard. Direct (conjugated) bilirubin produces immediate color with diazo reagent, and the indirect (conjugated) bilirubin produces color with diazo reagent only in the presence of 50% alcohol.

24.5 Procedure

$$NaNO_2 + HCl \longrightarrow NaCl + HNO_2$$

$$HO_3S-\langle\rangle-NH_2 + H_3O^+ \longrightarrow HO_3S-\langle\rangle-\overset{+}{N}H_3 + H_2O$$

Sulphanilic acid

$$HO_3S-\langle\rangle-\overset{+}{N}H_3 + HNO_2 \xrightarrow{-2H_2O} HO_3S-\langle\rangle-\overset{+}{N}\equiv N \longleftrightarrow HO_3S-\langle\rangle-N=\overset{+}{N}$$

Nitrous acid   Dizonium ion

Conjugated bilirubin + Diazotised sulphanilic acid → 2 Azobilirubin molecules

24.4 Reagents

1. **Diazo reagent**
   Sol. A – Dissolve 100 mg sulfanilic acid in 100 ml of 1.5% HCl.
   Sol. B – 0.5% aqueous solution of sodium nitrite. Diazo reagent is prepared by mixing 10 ml of solution A with 0.3 ml of solution B. Prepare just before use.
2. **Diazo Blank**: 1.5% v/v aqueous solution of HCl.
3. **Bilirubin standard (10 mg %)**: Dissolve 10 mg of bilirubin powder in 5 ml of 0.1 N NaOH, and add 80 ml of 2% BSA solution in saline. Then add 5 ml of 0.1 N HCl, and mix and make the volume 100 ml with 2% BSA solution. Store in brown bottle at 4°C.

24.5 Procedure

1. Take a set of 12 test tubes in duplicate and label as blank, standard (S1-S7), total (T) total control (Tc), direct (D), and direct control (Dc). Also take another six tubes for standard controls.
2. Add 20, 40, 60, 80, 100, 150, and 200 μl of 10 mg % bilirubin standard into the tubes labelled as S1-S7, respectively, and standard controls.
3. Pipette out 0.1 ml of serum to all the tubes labelled as T, Tc, D, and Dc. Make volume to 2 ml with distilled water in each tube except D and Dc. Add 4.9 ml distilled water to the tubes labelled as D and Dc.
4. Add 0.5 ml diazo reagent to all the tubes except blank, Tc and Dc, and standard controls. 0.5 ml of diazo blank was added to the tubes blank, Tc, Dc, and standard controls.
5. Add 3 ml of methanol to all the tubes except D and Dc. Mix and keep in dark for 30 min at room temperature and read absorbance at 540 nm.

Reagents	Blank	S_1	S_2	S_3	S_4	S_5	S_6	S_7	Total		Direct	
									T	Tc	D	Dc
Bilirubin standard (μl)	–	20	40	60	80	100	150	200	–		–	
Bilirubin amount (μg)	–	2	4	6	8	10	15	20				
Serum/urine (ml)	–	–	–	–	–	–	–	–	0.1	0.1	0.1	0.1
Distilled water (ml)	2.0	1.98	1.96	1.94	1.92	1.90	1.85	1.80	1.9	1.9	4.9	4.9
Diazo reagent (ml)	–	0.5	0.5	0.5	0.5	0.5	0.5	0.5	0.5	–	0.5	–
Diazo blank (ml)	0.5	–	–	–	–	–	–	–	–	0.5	–	0.5
Methanol (ml)	3	3	3	3	3	3	3	3	3	3	–	–

Mix well and keep in dark for 30 min at room temperature, read absorbance at 540 nm

24.6 Calculations

Plot a graph by taking bilirubin amount at x-axis vs absorbance at y-axis, and extrapolate bilirubin concentration of the given serum sample from the graph. The bilirubin concentration can also be calculated by applying equation:

$$\text{mg of bilirubin* in blood/dl} = \frac{(\text{OD of test} - \text{control}) \times \text{amount of standard} \times 100}{(\text{OD of standard} - \text{control}) \times \text{volume of sample (ml)} \times 1000**} = x \text{ mg/dl}$$

*Calculate total and direct bilirubin; **To convert μg to mg since amount of bilirubin is in μg in the volume of standard added to tubes.

24.7 Clinical Significance

Normal serum contains very low concentration of direct bilirubin, i.e., 0–0.2 mg/dl, and the total bilirubin concentration ranges from 0.2 to 1.0 mg/dl. Bilirubin analysis is a part of hepatic function test. Increased bilirubin concentration in blood causes jaundice. The estimation of direct (conjugated) and indirect (unconjugated) bilirubin may help in differential diagnosis of various types of jaundices – prehepatic (hemolytic), hepatic, and posthepatic (obstructive).

1. In prehepatic or hemolytic jaundice, the unconjugated bilirubin is increased due to increased destruction of red blood cells, and the liver is unable to cope with the increased demand for conjugation. In pernicious and chronic hemolytic anemia, the serum bilirubin rarely exceeds 3 mg/dl but may also reach up to 10 mg/dl.

24.7 Clinical Significance

2. In hepatic jaundice, conjugated bilirubin decreases along with decreased excretion of bilirubin. The concentration of total bilirubin increases greatly in blood without any significant change in their ratio.
3. In posthepatic jaundice (obstructive jaundice), the concentration of conjugated bilirubin is increased (up to 20 mg/dl) compared to unconjugated bilirubin. Increased serum bilirubin indicates extrahepatic biliary tract obstruction and hemolytic diseases.

Total serum bilirubin amount >40 mg/dl indicates hepatocellular obstruction and not extrahepatic obstruction. Urine bilirubin and urobilinogen are important in differential diagnosis of jaundice. Bilirubinuria is present in obstructive jaundice but absent in hemolytic jaundice.

To Determine Alanine and Aspartate Transaminase Activity in Serum

25.1 Theory

Transaminases are tissue-specific intracellular enzymes, which catalyze reversible transfer of α-amino group from amino acid to α-keto acid. They are present in almost all cells, but higher amounts occur mainly in the liver, brain, heart, and kidney. Two clinically important transaminases are serum glutamine oxaloacetate transaminase (SGOT), also called aspartate aminotransferase (AST), and serum glutamine pyruvate transaminase (SGPT) also called alanine aminotransferase (ALT). Alanine transaminase is present in high amount in the cardiac muscles but is also present significantly in other tissues of the body. ALT shows its high concentration in the liver and kidney, and trace amounts are present in the skin, pancreas, spleen, lungs, and cardiac and skeletal muscle. Both AST and ALT enzymes are not excreted in urine unless a kidney lesion is present. The measurement of these enzymes is a useful diagnosis indicator of liver functions.

25.2 Specimen Requirements

Serum sample is used which is stable for more than 24 h at 4°C. Hemolysis significantly increases AST levels but moderately increases ALT. Activity in erythrocytes is almost 6 times higher than the activity in serum. ALT shows higher day-to-day variation than AST. Its levels are higher in afternoon than morning.

25.3 Principle

The oxaloacetate and pyruvate formed by the action of AST and ALT from appropriate substrates couple with dinitrophenyl hydrazine (DNPH) to form a golden brown or red color in alkaline medium created with NaOH. The intensity of color is proportional to the enzyme activity which is measured at 540 nm and compared with pyruvic acid standard.

$$\underset{\text{Oxaloacetate}}{\begin{array}{c}COO^-\\|\\C=O\\|\\CH_2\\|\\COO^-\end{array}} + \underset{\text{2,4-DNPH}}{H_2N-N\overset{H}{\underset{}{}}\text{-}C_6H_3(NO_2)_2} \longrightarrow \underset{\text{Oxaloacetate – 2,4-DNPH}}{\begin{array}{c}COO^-\\|\\C=N-N\overset{H}{\underset{}{}}\text{-}C_6H_3(NO_2)_2\\|\\CH_2\\|\\COO^-\end{array}} + H_2O \xrightarrow{\text{Alkaline condition}} \text{Golden brown colour}$$

$$\underset{\text{Pyruvate}}{\begin{array}{c}COO^-\\|\\C=O\\|\\CH_3\end{array}} + \underset{\text{2,4-DNPH}}{H_2N-N\overset{H}{\underset{}{}}\text{-}C_6H_3(NO_2)_2} \longrightarrow \underset{\text{Pyruvate – 2,4-DNPH}}{\begin{array}{c}COO^-\\|\\C=N-N\overset{H}{\underset{}{}}\text{-}C_6H_3(NO_2)_2\\|\\CH_3\end{array}} \xrightarrow{\text{Alkaline condition}} \text{Golden brown colour}$$

25.4 Reagents

1. **Phosphate buffer (0.1 M, pH 7.4)**: Prepare phosphate buffer by dissolving 11.3 g Na_2HPO_4 and 2.7 g KH_2PO_4 in final volume of 1 l distilled water. Adjust pH to 7.4 using either KH_2PO_4 or Na_2HPO_4.
2. **Buffer substrate for AST**: Dissolve 1.33 g of L-aspartic acid and 30 mg α-ketoglutarate in about 20 ml of phosphate buffer. Add 3–5 pellets of NaOH to dissolve it and adjust pH to 7.4 by NaOH or HCl. Make volume up to 100 ml with phosphate buffer and store at 4°C.
3. **Buffer substrate for ALT**: Dissolve 0.9 g of alanine and 30 mg α-ketoglutarate in about 20 ml of phosphate buffer (pellets of NaOH may be added to dissolve alanine). Make the volume up to 100 ml with phosphate buffer, and adjust pH to 7.4 if required.
4. **Pyruvate standard (10 mM, pH 7.4)**: Dissolve 110 mg of sodium pyruvate in phosphate buffer, and make the volume up to 100 ml. Store working standard in small aliquots in a freezer, and one aliquot of standard may be used for preparing a calibration graph.
5. **Color reagent**: dissolve 20 mg of 2,4-DNPH in 100 ml of 1 N HCl, and store in brown bottle.
6. **0.4 M NaOH**

25.5 Procedure

1. Take a set of nine test tubes, and label them as blank, standard (S_1–S_6), test (T), and control (C).
2. Add 10, 20, 40, 60, 80, and 100 µl of pyruvic acid standard to the test tubes marked as S_1 to S_6, respectively.
3. Make volume in each tube to 100 µl using phosphate buffer.
4. Add 0.5 ml of buffered substrate to all the tubes. Then mix and incubate for 3 min at 37°C.
5. Add 100 µl of serum to test tube labelled as T. Mix and note time.
6. Incubate the tubes at 37 °C for 30 min for ALT and 1 h for AST. Keep at room temperature for 20 min.
7. Immediately after adding DNPH, add 100 µl of serum in control tube.
8. Add 5 ml of 0.4 N NaOH to all tubes, and mix and keep at room temperature for 10 min.
9. Read absorbance at 540 nm against blank.

Note For AST estimation the process is the same except the following:

1. Use AST substrate
2. Incubate for 60 min

Reagents	Blank	S_1	S_2	S_3	S_4	S_5	S_6	T	C
Pyruvic acid standard (µl)	–	10	20	40	60	80	100	–	–
Pyruvate amount (mmol)	–	0.1	0.2	0.4	0.6	0.8	1	–	–
Phosphate buffer (µl)	100	90	80	60	40	20	–	–	–
Buffered substrate for ALT/AST (ml)	0.5	0.5	0.5	0.5	0.5	0.5	0.5	0.5	0.5
Mix and incubate at 37°C for 3 min									
Serum (ml)	–	–	–	–	–	–	–	0.1	–
Mix well and incubate at 37°C for 30 min in case of ALT and for 1 h in case of AST									
DNPH (ml)	0.5	0.5	0.5	0.5	0.5	0.5	0.5	0.5	0.5
Serum (ml)	–	–	–	–	–	–	–	–	0.1
Mix well and incubate at 37°C for 20 min									
NaOH (ml)	5	5	5	5	5	5	5	5	5
Mix well and incubate at 37°C for 10 min, then read absorbance at 540 nm									

25.6 Calculation

To measure AST and ALT activities, pyruvate is used as the standard. In theoretical terms, for AST assay oxaloacetate should be used as the standard, and pyruvate should be used as the standard for ALT estimation. Since, the oxaloacetate formed during AST estimation is unstable and gets converted into pyruvate immediately; hence, pyruvate is preferred as the standard for AST estimation. Extrapolate pyruvate amount from graph and calculate activity. Enzyme activity is expressed as IU. One IU of enzyme activity is equal to 1 mmol of pyruvate/oxaloacetate formed/min/liter of serum at 37°C. The amount of pyruvate formed equals to enzyme activity. So, amount of pyruvate formed in m mol/min/l.

$$= \frac{\text{OD of (test} - \text{control)} \times \text{amount of standard (mmol)} \times 1000}{\text{OD of standard} \times \text{sample volume (ml)} \times \text{incubation time}}$$
$$= x \, \text{IU/L}$$

25.7 Clinical Significance

The normal range of ALT is 3–40 IU/L and AST is 5–45 IU/L in serum. Normally, the serum transaminase levels are low, but the extensive tissue damage enhances their serum levels. The AST activity in serum is increased in myocardial infarction after 20–36 h of onset, and hence measurement of AST is used in the diagnosis of myocardial infarction. Increased activities are also observed in acute pancreatitis, crushed muscle injuries, viral hepatitis, and other forms of liver diseases associated with liver necrosis. Peak values of transaminases are seen between 7th and 12th days of infection and return toward normal by 3rd week if recovery is uneventful.

ALT is generally higher than AST in infectious hepatitis and other inflammatory conditions of liver or hepatic tissue damage. In such conditions, ALT/AST ratio becomes greater than unity which is less than one in normal conditions. ALT is more liver-specific enzyme than AST. AST levels also show increase in progressive muscular dystrophy and dermatomyositis. ALT and AST levels are increased in viral hepatitis (10–100 times of normal value). The course of liver damage in a patient may be monitored by successive determination of serum AST and ALT activities.

To Estimate the Activity of Alkaline Phosphatase in Serum

26.1 Theory

Alkaline phosphatase (ALP) is a membrane-bound enzyme present mainly in the liver, bone, intestine, and placenta. ALP shows maximum activity at about pH 10. This enzyme is intracellular and only a small amount is present in plasma. In certain disease conditions, the breakdown of cell causes release of enzyme to plasma leading to increased enzyme levels in plasma. The ALP of normal serum in adults is mainly derived from the liver and bone and small amount from intestinal component. During childhood the majority of alkaline phosphatase is of skeletal origin. During pregnancy, ALP is also contributed from the placenta. ALP is activated by magnesium ions. The activity in serum is due to isoenzymes from various organs, but the major contribution occurs from the liver.

26.2 Specimen Requirements

Serum sample is used. Refrigerated sample is stable for 2–3 days if stored at 0–4°C and for 1 month at −25°C. EDTA or oxalate anticoagulants cause extensive inhibition. Half-life of ALP is 7–10 days.

26.3 Principle

Alkaline phosphatase at alkaline pH hydrolyzes phenyl phosphate to phenol. Phenol condenses with 4-amino-antipyrin. The condensed product is oxidized by alkaline potassium ferricyanide to give red-colored complex which is measured at 520 nm.

Phenyl phosphate $\xrightarrow{\text{ALP, pH 10}}$ Phenol + Phosphate

Phenol + 4-aminoantipyrine $\xrightarrow{\text{Potassium ferricyanide}}$ Orange red complex

26.4 Reagents

1. **Disodium phenyl phosphate (0.1 M)**: Dissolve 2.18 g of disodium phenyl phosphate in distilled water, and make final volume to 100 ml. Store at 4°C.
2. **Na_2CO_3-$NaHCO_3$ buffer (pH 10.0)**: Prepare 100 ml 0.1 M Na_2CO_3-$NaHCO_3$ buffer, (pH 10) by adding 51 ml 0.1 M Na_2CO_3 to 49 ml 0.1 M $NaHCO_3$, and adjust pH if required. Also prepare 0.5 N bicarbonate buffer separately.
3. **0.5 N NaOH**.
4. **4-aminoantipyrine (0.6%) aqueous solution**. Store at 4°C.
5. **Potassium ferricyanide (2.4%)**. Store at 4°C.
6. **Stock phenol standard (100 mg %) in 0.1 N NaOH**. Store at 4°C.
7. **Working phenol standard (10 mg %)**. Prepare in distilled water.

26.5 Procedure

1. Take a set of test tubes in duplicate and mark them as blank, standard (S_1 to S_6), control (C), and test (T).
2. Add working phenol standard according to table in tubes S_1 to S_6. Add 1 ml each of carbonate-bicarbonate buffer and substrate in all tubes. Incubate for 5 min.
3. Then add 0.1 ml serum in test sample and add 1 ml distilled water in blank. Incubate for 15 min.
4. Then add 1 ml NaOH to stop reaction followed by 0.1 ml serum in control tube, 1 ml of bicarbonate, 1 ml 4-aminoantipyrine, and 1 ml potassium ferricyanide.

Reagents	Blank	S_1	S_2	S_3	S_4	S_5	S_6	T	C
Working phenol standard (ml)	–	0.1	0.2	0.4	0.6	0.8	1	–	–
Amount of phenol in added volume (μg)	–	1	2	4	6	8	10	–	–
Carbonate/bicarbonate buffer (ml)	1	0.9	0.8	0.6	0.4	0.2	–	0.9	0.9
Disodium phenyl phosphate (ml)	–	1	1	1	1	1	1	1	1
Incubate at 37°C for 5 min									
Serum (ml)	–	–	–	–	–	–	–	0.1	–
Distilled water (ml)	1	–	–	–	–	–	–	–	–
Mix well and incubate at 37°C for 15 min									
0.5 N NaOH (ml)	1	1	1	1	1	1	1	1	1
Serum (ml)	–	–	–	–	–	–	–	–	0.1
0.5 N bicarbonate buffer (ml)	1	1	1	1	1	1	1	1	1
4-aminoantipyrine (ml)	1	1	1	1	1	1	1	1	1

(continued)

Potassium ferricyanide (ml)	1	1	1	1	1	1	1	1	1
Mix and read the absorbance at 520 nm									

26.6 Calculations

The enzyme activity is calculated in terms of King-Armstrong unit (KAU). One KAU is defined as 1 U of ALP as that liberating 1 mg of phenol from substrate/15 min/100 ml serum. Plot the graph between OD and the amount of phenol. The activity can also be calculated as below:

$$\text{ALP activity} = \frac{\text{OD of (test} - \text{control)} \times \text{amount of phenol} (\mu g) \times 100}{\text{OD of standard} \times \text{volume of sample (ml)} \times 1000}$$
$$= x \text{ KAU/dl or } x \text{ U/dl}$$

26.7 Clinical Significance

Normal serum ALP levels in adults are 3–14 U/dl or 30–120 U/L. The tissues such as the liver, bone, and placenta possess very high amounts of ALP activity. Increase in serum ALP activity is usually associated with hepatobiliary obstruction (all forms of jaundice except hemolytic jaundice). In hepatobiliary disease, the concentration of ALP is increased due to defect in excretion of ALP through bile. ALP mainly produced in bone tissue reaches to the liver and excreted through bile. Increased ALP levels are also observed in increased osteoblastic activity, rickets, hyperthyroidism, and bone disorders. A small increase in ALP activity may also occur in congestive heart failure, intra-abdominal bacterial infections, and intestinal diseases. Small increase may also be observed in multiple myeloma and impaired calcium absorption. The elevation of ALP is more marked in extrahepatic obstruction (gallstone, cancer of the head of the pancreas, etc.) than intrahepatic obstruction. Decrease in serum ALP is observed during severe anemia, celiac disease, zinc and magnesium deficiency, scurvy, cretinism, hypothyroidism, and hypophosphatemia.

To Estimate the Activity of Acid Phosphatase in Serum

27.1 Theory

The greatest activity of acid phosphatase (ACP) occurs in semen (prostate gland). ACP is lysosomal enzyme. The prostatic enzyme is found in lysosome of prostrate epithelium. Small amounts are also present in the platelets, leucocytes, liver, kidney, spleen, and pancreas. Approximately one-half of the total serum ACP activity in males is contributed from the prostate. The rest of activity is due to the liver and disintegrating platelets and erythrocytes. In females, activity is from the liver, erythrocytes, and platelets.

27.2 Specimen Requirements

Serum is used. Separate serum and analyze it immediately. Avoid hemolysis. Citrate buffer or acetic acid may be added to attain pH of 5.5–6.5 to preserve activity.

27.3 Principle

Principle is the same as for ALP except that the substrate is hydrolyzed by enzyme ACP at pH 5.0 with liberation of phenol.

27.4 Reagents

Same as used for ALP activity assay except the use of citrate buffer instead of carbonate-bicarbonate buffer.

Citrate Buffer (pH 5.0): Dissolve 21 g of citric acid and 7.5 g NaOH in about 450 ml distilled water, and adjust pH 5.0 and make final volume to 500 ml with distilled water.

27.5 Procedure

Procedure is the same as explained for ALP activity except the use of citrate buffer instead of carbonate-bicarbonate buffer.

27.6 Calculations

The enzyme activity is calculated in terms of KAU as explained for ALP. Plot a graph between OD and the amount of phenol. Activity can also be calculated using equation explained for ALP.

27.7 Clinical Significance

Normal range of ACP is 1–3.5 KAU/dl. ACP levels increase 40–50 times of normal levels in prostate cancer. Earlier, ACP was used for diagnosis of prostate cancer that is now replaced by prostate-specific antigen. Slight increase in ACP activity is also observed in hyperparathyroidism with skeletal involvement, malignant inversion of bones by cancers such as female breast cancer, and Paget's disease. Non-prostrate ACP fraction increases in myelocyte leukemia and some other hematological disorders.

To Determine Serum and Urinary Amylase Activity

28.1 Theory

Amylases are a group of hydrolyzing enzyme which hydrolyzes starch and glycogen. There are two types of amylases: α-amylase (hydrolyzes randomly α 1–4 glycoside links, also known as endoamylases, e.g., human α-amylase) and β-amylase (hydrolyzes β 1–4 glycosidic linkage). Salivary glands, pancreas, and fallopian tubes show highest amylase activities, while ovaries, small and large intestine, and skeletal muscle possess low activity of amylase.

28.2 Specimen Requirements

Serum is used which is stable for 7 days at room temperature and for 1 month at 4°C. For the blood test, patients should not eat or drink anything except water for 2 h before the test. For the urine test, patients should drink enough fluids during the 24 h test to avoid dehydration. Collect 24 h urine or 1 h urine in container with preservative. Acidic urine makes amylase unstable; hence pH is adjusted to alkaline range before storage at 4°C.

28.3 Principle

Iodine gives blue color with starch and its degradation products containing longer chain linear polysaccharides. As the chain length shortens to less than 45 units, the blue color changes through purple to red and brown until with less than 12 units no color is observed. Amylase hydrolyzes starch to glucose and maltose; the time required for the complete digestion of a certain amount of starch is determined by periodic testing with iodine and, this time, is converted to Somogyi units.

28.4 Reagents

1. **Starch substrate solution**: Dissolve 75 mg of starch in about 20 ml of distilled water with the aid of heat. Add 250 mg of NaCl, dissolve, and make the final volume 100 ml with distilled water.
2. **Stock iodine solution (0.1 M)**: Dissolve 40 g of potassium iodide in about 100 ml distilled water, and add 12.7 g of iodine. Mix well to dissolve, and make the final volume 1000 ml with distilled water.
3. **Working iodine solution (0.05 M)**: 5 ml of stock iodine solution is diluted to 100 ml with distilled water. Prepare fresh before use. Store in a brown bottle.

28.5 Procedure

Pipette out 2 ml starch substrate in a test tube, and keep at 37°C in water bath for 3–5 min. Add 0.5 ml serum and mix and then note time. After 5 min, remove 0.2 ml of the above mixture from the incubator and add to 0.2 ml of working iodine solution kept outside and mix and observe the color of iodine solution. If the color of iodine solution is purple brown, this indicates that the starch present in that tube is not completely hydrolyzed by the action of amylase enzyme. Again remove 0.2 ml of starch serum mixture at the end of sixth min of incubation time, and add 0.2 ml iodine solution. The procedure is repeated at every 1 min interval until the iodine solution gives yellow color with serum starch mixture. Note the time for calculating enzyme activity.

Urinary Amylase (Diastase) Estimation Procedure for the estimation of urinary amylase is same as that of serum amylase. Report the amylase activity in terms of per hour urine sample.

Note If the activity of urinary amylase and serum amylase is very high, dilute the urine and serum and repeat the test and calculate the value and multiply the value obtained with dilution factor.

28.6 Calculation

One Somogyi unit (SU) is defined as 1.5 mg of starch (2 ml substrate in present case) hydrolyzed by the action of amylase present in 100 ml serum. If time taken for hydrolysis is "t" min, then amylase activity (SU/100 ml serum) can be calculated as

$$= \frac{8 \times 100}{t \times \text{vol of sample used}}$$

28.7 Clinical Significance

Normal amylase activity ranges from 80 to 180 SU/dl in serum and from 1500 to 1800 SU/24 h urine specimens. In acute pancreatitis, enzymes from damaged cells may pass into the blood in considerable amount and more amount of amylase excreted through urine. In renal failure, the clearance of amylase is reduced and serum amylase increased, while urinary amylase excretion is normal.

In acute pancreatitis, the serum amylase value reaches from 1000 to 6000 SU/dl at 24–48 h after onset of the attack with a concomitant large increase in urinary amylase. The increase in activity starts within an hour of onset of pain and usually returns to normal in 4–8 days being partly removed by renal excretion. In chronic pancreatitis and later stage of carcinoma of pancreas, increased levels of serum and urinary amylase are observed. Increase of amylase over 1000 SU/dl occurs in biliary tract disease without pancreatitis. Increase in amylase activity also occurs in intestinal obstruction and perforated peptic ulcer.

To Estimate the Activity of Lipase in Serum

29.1 Theory

Lipase is the enzymes that cause hydrolysis of glycerol esters of long chain fatty acids into glycerol and free fatty acids. Only the ester bond at carbons 1 and 3 (α-positions) is attacked, and the reaction produces two moles of fatty acid and one mole of 2-acylglycerol (β-monoglyceride) per mole of substrate. The latter is resistant to hydrolysis, but it can spontaneously isomerizes to the 3-acylglycerol, that is, α-form. This isomerization splits off the third fatty acid, a process that occurs at a much slower rate. Lipase acts at an oil-water interface on emulsified substrates. Most of the serum lipase is produced in pancreas, but pulmonary, gastric, and intestinal mucosa also secretes lipase. Lipase activity can also be demonstrated in leucocytes in adipose tissue cells and in milk.

29.2 Specimen Requirements

Serum sample is used which is stable for several days at room temperature. Store refrigerated to avoid bacterial contamination.

29.3 Principle

Fatty acids released by the action of lipase on substrate for a period of 24 h at 37 °C (pH 7.4) are determined by titrating against standard solutions of NaOH using phenolphthalein as indicator.

29.4 Reagents

1. **Olive oil emulsion (50%)**: Take equal parts of olive oil and 5% aqueous solution of gum acacia and homogenize it.
2. **Phosphate buffer, 0.1 M, pH 7.4.**
3. **Ethanol.**
4. **0.05 M NaOH solution.**
5. **Phenolphthalein (1%) alcoholic solution.**

29.5 Procedure

1. Take two test tubes and label as test and control.
2. Add 0.4 ml of olive oil emulsion and 0.2 ml serum and 0.5 ml phosphate buffer to both the tubes.
3. Vortex the tubes and incubate the tube labelled as test at 37°C for 24 h and control tube at 4°C for 24 h.
4. Take out both the tubes and add 3 ml of ethanol to test and control tubes.
5. Vortex and add two drops of phenolphthalein to both the tubes, mix, and titrate both the tubes against 0.05 N NaOH solution until a pink color is observed. Note the volume of NaOH used.

29.6 Calculation

One Cherry Crandall Unit (CCU) is defined as 1 ml of 0.05 N NaOH required to neutralize the free fatty acid liberated by the action of lipase on olive oil substrate at 37°C (pH 7.4), for a period of 24 h by 1 ml serum.

$$\text{Lipase activity/ml serum} = \frac{\text{titration value} (T - C) \times 1}{\text{Volume of sample}}$$
$$= x \, \text{CCU/ml}$$

29.7 Clinical Significance

Measurement of lipase activity in serum, plasma, or ascetic fluid is used exclusively for the diagnosis of pancreatitis. Normal level of lipase in serum is up to 0–160 U/L. In acute pancreatitis the serum lipase activity rises within 2–10 h. Peak level at about 12 h and the value may return to normal within 48–72 h. In acute pancreatitis, the lipase value increases 2–50-fold than normal. The increase in lipase activity is also observed in obstruction of pancreatic duct by calculi or carcinoma of the pancreas. In acute and chronic renal disease, there is increase in serum lipase activity.

Qualitative Analysis of Ketone Bodies in Urine

30.1 Theory

The three compounds, namely, acetone, acetoacetic acid, and β-hydroxybutyrate, are called ketone bodies. Ketone bodies are water-soluble and energy-yielding substances. The synthesis of ketone bodies is called ketogenesis. The ketogenesis process occurs in the liver particularly during periods of low food intake, carbohydrate-restrictive diets, starvation, or in untreated type 1 diabetes mellitus. Ketone bodies are transported from the liver to the extrahepatic tissues and converted into acetyl-CoA which then enters the citric acid cycle and is utilized for energy production (Fig. 30.1). The liver, however, is unable to metabolize ketone bodies. The tissues which lack mitochondria, i.e., erythrocytes, also cannot utilize ketone bodies. Ketone bodies are major fuel source for the brain during starvation and can meet 50–70% of total energy needs. The relative proportion of ketone bodies present in blood may vary from 78% (β-hydroxybutyric acid) to 20% (acetoacetic acid) and 2% (acetone). Tests for ketone bodies should be done on fresh urine sample.

Ketone bodies

30 Qualitative Analysis of Ketone Bodies in Urine

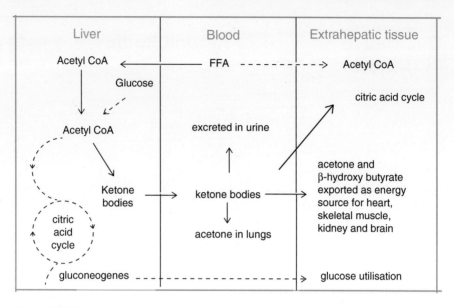

Fig. 30.1 Utilization of ketone bodies

30.2 Rothera's Test for Acetoacetic Acid and Acetone

30.2.1 Principle

Alkaline nitroprusside reacts with keto-group of acetone and acetoacetic acid to form a purple-colored complex.

30.2.2 Reagents

(a) **Rothera's reagent: dry mixture**

Take ammonium sulfate and sodium nitroprusside in 100:1 ratio. Grind well to mix powder of salts.

(b) **Conc. ammonia** in liquid form

30.2.3 Procedure

5 ml of urine is saturated with Rothera's reagent in a test tube. Then 0.5–1.0 ml of conc. ammonia is added through the sides of the tube in such a way that it layers on top of the urine. Any change in color is observed within 30–60 s.

30.2.4 Result

Appearance of a purple/red ring at the junction of the two layers within 30–60 s indicates the presence of acetone and diacetic acid. Based on the intensity of the color formed, the result can be represented as grade from trace to 3^+.

The presence of L-dopa and phenyl pyruvic acid in urine may produce weak false positive reactions. In case of suspicion of a false positive test, the urine sample is heated in a test tube for 1 min, allowed to cool, and Rothera's test is repeated. The heated urine will not produce a positive Rothera's test due to ketone bodies.

30.3 Gerhardt's $FeCl_3$ Test for Acetoacetic Acid

This test is based on the reaction of $FeCl_3$ with acetoacetate producing a red wine color. The test is non-specific, and antipyrine gives similar color.

30.3.1 Reagent

10% aqueous solution of $FeCl_3$.

30.3.2 Procedure

Take 5 ml of urine and add 10% $FeCl_3$ solution drop by drop. Red wine color (purple) indicates positive test.

30.4 Ketostix Test for Acetone and Acetoacetate

Ketostix are plastic strips, one end of which is impregnated with sodium nitroprusside and glycine. When these strips are immersed in urine, sodium nitroprusside and glycine react with acetone and acetoacetate to give lavender or purple color.

30.4.1 Procedure

Dip the test end of strip in fresh specimen of urine, and remove immediately, briefly touching the tip on side of container to remove excess liquid. Compare color with chart. Ketostix is sensitive to 10 mg/dl acetoacetic acid and 25 mg/dl acetone.

30.5 Detection of β-Hydroxybutyrate

30.5.1 Principle

β-hydroxybutyrate forms acetoacetate on oxidation with H_2O_2 which can be detected with Rothera's test.

30.5.2 Procedure

Add few drops of acetic acid in about 10 ml of urine diluted 1:1 with distilled water. Boil for few min to remove acetone and acetoacetic acid. Add about 1 ml of H_2O_2, warm gently, and apply Rothera's test.

Note If Rothera's test is positive before oxidation with H_2O_2 (with fresh urine sample), then acetone and acetoacetic acid ketone bodies are present. If Rothera's test give positive result after oxidation with H_2O_2, then *β*-hydroxybutyrate is present in urine.

30.6 Clinical Significance

In normal conditions, the ketone bodies are produced constantly by the liver and utilized by peripheral tissues. The normal concentration of ketone bodies in blood is about 1–3 mg/dl, and their excretion in urine is negligible, that is, undetectable by routine urine tests. The concentration of ketone bodies increases in the blood when the rate of synthesis exceeds the rate of utilization, a condition named ketonemia. The higher increase of ketone bodies in blood leads to their excretion in urine. The excretion of ketone bodies in urine is called ketonuria. In untreated diabetes patient and starving patient, the serum ketone body levels may reach 90 mg or above per 100 ml of serum, and urinary excretion may reach 5000 mg/24 h urine. In type I diabetes, low insulin levels impair carbohydrate metabolism that leads to accumulation of acetyl CoA and its conversion to ketone bodies. During starvation also increased lipolysis causes overproduction of acetyl CoA which is diverted for overproduction of ketone bodies. Both these conditions result in very high levels of ketone bodies, lowering the blood pH, and kidneys excrete very acidic urine.

Qualitative Test for Bile Pigments and Urobilinogen in Urine

31.1 Theory

Bile pigments are breakdown products of the blood pigment hemoglobin that are excreted in bile. Bilirubin (orange or yellow in color) and its oxidized form biliverdin (green) are two important bile pigments. Bilirubin is formed by the degradation of heme in reticuloendothelial system. After formation of unconjugated bilirubin, albumin transports it to the liver where it conjugates with glucuronic acid to form bilirubin diglucuronide. The conjugated bilirubin is excreted through bile. Unconjugated bilirubin is water insoluble, so it is unable to pass through the glomerulus and is therefore not found in urine, whereas conjugated bilirubin is water soluble and is filtered in the glomerulus and subsequently reabsorbed by the renal tubules; hence normally bile pigments are absent in urine. Bilirubin glucuronides are hydrolyzed by intestinal bacterial enzyme β-glucuronidases to liberate bilirubin which is converted to a colorless compound "urobilinogen." A portion of the urobilinogen formed is reabsorbed by enterohepatic circulation. Urobilinogen can be converted to a yellow-colored compound "urobilin" in the kidneys and excreted. The major part of intestinal urobilinogen is reduced to stercobilin by bacterial action which imparts brown color to feces and is excreted (Fig. 31.1).

31.2 Tests for Bile Pigments in Urine

31.2.1 Fouchet's Test

31.2.1.1 Principle
The sulfate radicals present in urine react with $BaCl_2$ to form precipitates of barium sulfate. Bilirubin present in the urine gets absorbed on the precipitates, and hence it get conc. on treatment with Fouchet's reagent. $FeCl_3$ in the presence of TCA oxidize the bilirubin to biliverdin and produce blue or green color.

Fig. 31.1 Bilirubin metabolism

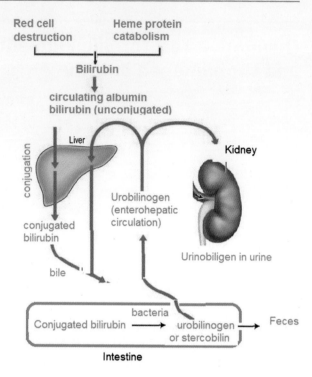

31.2.1.2 Reagents
(a) **33% v/v aqueous solution of acetic acid**.
(b) **10% aqueous solution of $BaCl_2$**.
(c) **Fouchet's reagent**: Dissolve 5 g TCA in about 60 ml of distilled water and add to it 10 ml of 10% $FeCl_3$ solution, mix, and make the final volume 100 ml with distilled water.

31.2.1.3 Procedure
Acidify urine sample with few drops of 33% acetic acid. To 10 ml of acidified urine sample, add 5 ml of 10% $BaCl_2$ solution. Mix well, and the precipitate is filtered through Whatman No. 1 filter paper. Dry the precipitates in between two layers of filter paper. Unfold the filter paper and add one or two drops of Fouchet's reagent on the precipitates and observe the color produced. Green-blue color formed on the precipitate indicates bile pigments in the urine.

31.2.2 Hunter's Test

31.2.2.1 Principle
Bilirubin present in the urine sample gets adsorbed on barium sulfate precipitates. This bilirubin reacts with diazotized sulfanilic acid to form pink-red color azobilirubin.

31.2.2.2 Reagents
1. **Diazo reagent**:
   Sol. A: Dissolve 0.1 g of sulfanilic acid in 1.5 ml conc. HCl, and make the final volume 100 ml with distilled water.
   Sol. B: 0.5% aqueous solution of sodium nitrite.
   Mix 10 ml solution A with 0.3 ml solution B just before use; this is diazo reagent.
2. **10% aqueous solution of $BaCl_2$**
3. **Absolute alcohol**
4. **6% Na_2HPO_4**

31.2.2.3 Procedure
To 5 ml of urine, add 2 ml of 10% $BaCl_2$ solution. Mix well and centrifuge at 3000 rpm for 5 min. Discard the supernatant and add 2 ml of distilled water to the pellet, mix, and centrifuge at 3000 rpm for 5 min for washing the precipitates. Discard the supernatant, and add 0.5 ml of diazo reagent. Mix and add 2 ml of ethanol and 0.5 ml of 6% Na_2HPO_4, and observe the color. Purple-red color indicates the presence of bile pigments in urine.

31.2.3 Gmelin's Test

Add 3 ml of urine carefully over 5 ml conc. HNO_3 in a test tube. Wait for few min. The presence of bile pigments shows a play of colors from yellow, red, violet, blue to green due to formation of derivatives of bile pigments by action of HNO_3.

31.2.3.1 Clinical Significance
Normally bile salts are absent in urine. The color of urine may be dark yellow to greenish brown depending upon the concentration of bile pigments, and when shaken urine becomes foamy. Bilirubin is found in urine in posthepatic or extrahepatic obstructive jaundice and in those types of hepatic jaundice in which there is cholestasis. It is thus an indicator of regurgitation of conjugated bilirubin which being water soluble and loosely attached to plasma albumin passes into the urine. It is not present in urine in prehepatic (hemolytic) jaundice. The unconjugated bilirubin, circulating bound to albumin, is poorly soluble in water. The color of urine may be dark yellow to greenish brown depending upon the amount of bile pigments. It also forms foams readily when shaken.

31.3 Urobilinogen in Urine

31.3.1 Ehrlich's Test

31.3.1.1 Principle
A red-colored complex is formed when urobilinogen reacts with *p*-dimethyl amino benzaldehyde in conc. HCl. Sodium acetate is used to reduce the acidity after the reaction of urobilinogen with Ehrlich's reagent.

31.3.1.2 Reagents

Ehrlich's Reagent Dissolve 0.5 g para-dimethyl amino benzaldehyde in 100 ml of 150:100 ratio of conc. HCl and distilled water.

31.3.2 Saturated Sodium Acetate Solution

31.3.2.1 Procedure
Mix equal volume (5 ml each) of urine and Ehrlich's reagent, and keep for 10 min. Then add 5 ml of saturated sodium acetate solution, and observe the color produced. If red color is produced, urobilinogen is present in the sample. In case no color developed, warm at 60 °C.

Use fresh sample of urine; otherwise false negative result may be observed, as urobilinogen on standing is oxidized to urobilin. Urobilin shows negative result with Ehrlich's reagent. A false positive result may be given by natural substances like porphobilinogen and drugs like para-aminosalicylic acid, sulfonamides, and sulfonylureas. In order to distinguish between urobilinogen and porphobilinogen, add 5 ml chloroform to the red color solution formed in Ehrlich's test. Shake well and let the layer to separate. Red color in the lower organic layer indicates the presence of urobilinogen, while the color in the above aqueous layer indicates porphobilinogen.

31.4 Precautions

1. Use fresh urine sample.
2. If bilirubin is present in the sample, first remove the bile pigments by Fouchet's test or by adding 10% $BaCl_2$ to 5 volume of urine. Centrifuge the solution and take the supernatant and detect urobilinogen by Ehrlich's test.

31.5 Clinical Significance

Normally small amount of urobilinogen is excreted through urine (1–3.5 mg/24 h urine). In hemolytic jaundice, urobilinogen is raised but not high because the liver is not able to completely excrete the increased quantity and absorbed in the intestine. So, in hemolytic jaundice, positive test for urobilinogen and urobilin is obtained with negative test for bilirubin. In extrahepatic (obstructive) jaundice, the complete absence of bilirubin in the intestine results in the absence of urobilinogen in urine. In liver cirrhosis, in spite of the reduced amount of urobilinogen found in the intestine, there is often an appreciable increase in urine urobilinogen which may be of diagnostic value; as in these cases, bilirubin may not be present in the urine. Pink color is also given by porphobilirubin and is of significance in porphyrias. It may be distinguished from urobilinogen by the absence of color intensification on addition of sodium acetate.

Determination of Total Lactate Dehydrogenase Activity in Serum Sample

32.1 Theory

Lactate dehydrogenase (LDH) is an enzyme involved in glucose metabolism. LDH is widely distributed in all cells but especially abundant in cardiac and skeletal muscles, liver, kidney, and red blood cells. LDH contains five isoenzymes (LDH1, LDH2, LDH3, LDH4, and LDH5). These isoenzymes can be separated by electrophoresis. LDH1 has more positive charge and hence fastest in electrophoresis mobility, while LDH5 is slowest in mobility. LDH is an oligomeric enzyme made up of four polypeptide subunits. Two types of subunits M (for muscle) and H (for heart) are produced by different genes. LDH1 is predominantly found in the heart, while LDH5 occurs in skeletal muscles. Total LDH activity is measured as combined activity of all isoenzymes in plasma (LDH1 and LDH2 are predominant in plasma than LDH3, LDH4, and LDH5 which are present only in small amounts).

32.2 Specimen Requirements

Serum/plasma (heparin) is the preferred sample. Store at room temperature, do not refrigerate or freeze, and sample should be free from any clot.

32.3 Principle

LDH is an oxidoreductase enzyme that catalyzes the reversible oxidation of lactate to pyruvate using NAD^+ as hydrogen acceptor. The equilibrium favors the conversion of pyruvate to lactate at pH 7.4–7.8. The decrease in absorbance at 340 nm is directly proportional to LDH activity.

$$\text{Lactate} + \text{NAD}^+ \underset{}{\overset{\text{LDH}}{\rightleftarrows}} \text{pyruvate} + \text{NADH} + \text{H}^+$$

32.4 Reagents

1. Buffer substrate solution (consisting of 50 mM phosphate buffer, pH 7.5 and 0.6 mM pyruvate, and 0.095% sodium azide)
2. $NADH_2$: 0.18 mmol/L

32.5 Procedure

Take 2 ml of NADH solution and add 50 µl of serum sample. Incubate for 2 min, and then add 2 ml buffer substrate. Measure the decrease in absorbance at 340 nm at 1 min interval for 3 min. If absorbance change exceeds 0.1/min, dilute sample accordingly and include dilution factor in calculation.

32.6 Calculation

Lactate dehydrogenase activity/liter of serum

$$= \frac{\Delta\text{ OD/min} \times \text{total volume in cuvette} \times 1000}{(\text{molar extinction coefficient}) \times \text{volume of serum used (ml)}}$$
$$= \text{U/L}$$

Molar absorption coefficient of NADH at 340 nm = $6.22 \times 10^3 \times 10^{-6}$/µ mole/cm.

32.7 Clinical Significance

Normal total LDH activity is 100–190 U/L at 37°C. The isoenzymes of LDH have diagnostic value in heart and liver diseases. In a healthy individual, LDH2 value is higher than LDH1 in serum. In myocardial infarction, the peak levels of LDH1 activity is much higher than LDH2 (although both LDH1 and LDH2 may be elevated, but LDH1 is more elevated), and that occurs within 12–24 h after onset of infarction and elevated levels persist for 10–14 days. LDH5 levels increase predominantly in muscular dystrophy, while in liver infarction or liver malignancy both LDH4 and LDH5 are increased. It should be noted that LDH activity is 80–100 times more in red blood cells (RBC) than in serum. Hence for estimation of enzyme activity, serum should be free of hemolysis.

To Measure Activity of Creatine Kinase in Serum

33.1 Theory

Creatine kinase (CK), also called creatine phosphokinase, exists as three isoenzymes, i.e., CK_1, CK_2, and CK_3. Each isoenzyme is dimeric enzyme composed of subunits M (muscle) or B (brain). The isoenzyme CK_1 contains subunit BB and occurs primarily in the brain. CK_2 possess subunit MB and are found in the heart. Skeletal muscles primarily contain the MM isoform (CK_3). Healthy individuals typically contain the MM isoform and a small amount of the MB isoform in their serum. The enzyme is not found in the liver, kidney, and blood cells. Various conditions including skeletal muscle injury and myocardial damage release CK-MB into the bloodstream. In myocardial infarction, levels of both CK-total and CK-MB increase significantly, but CK-MB is considered a specific cardiac marker.

33.2 Sample Requirement

Serum or plasma (heparinized/EDTA) is used. Protect from light. Loss of activity occurs within 7 days at 4°C or within 24 h at 25°C.

33.3 Principle

Creatine kinase enzyme hydrolyzes creatine phosphate to liberate creatine and ATP at pH 7.4. Enzyme hexokinase converts glucose to glucose-6-phosphate in presence of ATP. Glucose-6-phosphate is converted to 6-phosphogluconolactone. The change in absorbance is monitored at 30 s intervals for 3 min at 340 nm. The enzyme in serum is relatively unstable and loses its activity due to sulfhydryl group oxidation at the active site of the enzyme. The enzyme activity is partially restored by incubating the enzyme reaction with sulfhydryl containing compounds such as

N-acetylcysteine, thioglycerol, dithiothreitol, cysteine, etc. The reaction sample is also added with an antibody specific to CK-M subunit which inhibits CK-M monomer.

$$\text{Creatine phosphate} + \text{ADP} \underset{}{\overset{CK}{\rightleftarrows}} \text{Creatine} + \text{ATP}$$

$$Glucose + \text{ATP} \underset{}{\overset{\text{Hexokinase}}{\rightleftarrows}} \text{Glucose-6-Phosphate} + \text{ADP}$$

$$\text{Glucose-6-Phosphate} + \text{NADP} \xrightarrow{\text{G-6-P dehydrogenase}} \text{6-phosphogluconolactone} + \text{NADPH}_2$$

33.4 Enzyme Reagents

Imidazole buffer (10 mM, pH 7.7)
Phosphocreatine (30 mM)
N-acetylcysteine (20 mM)
Magnesium acetate (10 mM)
Glucose (20 mM)
Glucose-6-phosphate dehydrogenase $\geq$ 1.5 KU/L
Hexokinase $\geq$ 2.5 KU/L
EDTA, ADP, and NADP (2 mM each)
AMP (5 mM)

33.5 Procedure

Keep enzyme reagent at 37°C before use. Take 1 ml of this enzyme reagent in a thermostated cuvette; add 0.05 ml of diluted serum/plasma, mix thoroughly, and incubate exactly for 5 min at 37°C and then read absorbance at 30 s intervals for further 3 min.

33.6 Calculation

CK-MB activity/L of serum (IU/L)

$$= \Delta \text{OD/min} \times 6752 \times \text{dilution factor}$$

where, factor 6752 is obtained as:

$$= \frac{\text{total volume in cuvette}}{(\text{molar extinction coefficient of NADPH}) \times \text{volume of serum used (ml)}}$$

$$= \frac{1.05}{6.22 \times 10^{-3} \times 0.05}$$

$$= 3376$$

Note If we use an antibody against CK-M, we get value of only CK-B from this enzyme. So to get activity of CK-MB, multiply value 3376 with factor 2 (i.e., $3376 \times 2 = 6752$). If we follow the above process without CK-M antibody, then value obtained will be for total CK activity. Calculation of CK-MB % is used to predict the occurrence of myocardial infarction. CK-MB % is calculated as:

$$\text{CK-MB\%} = \frac{\text{CK-MB IU/L}}{\text{Total CK IU/L}} \times 100$$

33.7 Clinical Significance

Normal range of CK-MB activity is 24–195 IU/L in males and 24–170 IU/L in females. The total CK activity range is 60–400 U/L in males and 40–150 U/L in females. The increase in CK activity is mainly caused by skeletal muscle disease, i.e., muscular dystrophy, myocardial infarction, and cerebrovascular damages. Other factors like polymyositis, viral myositis, and hypothyroidism also increase CK levels. The increase in the CK activity (particularly CK_2 or MB isoenzyme) is observed within 6–8 h after onset of myocardial infarction. The activity reaches to maximum levels after 12–24 h and comes to normal after 3–4 days. Increase in levels of both total CK activity and CK-MB activity can be observed in the serum of patients suffering myocardial infarction or after physical exercise. The increase is not observed in heart failure and coronary insufficiency. The CK-MB % less than 5.5% indicates probability of no myocardial infarction or myocardial infarction occurred in less than 4–6 h prior to sample collection. The CK-MB % between 5.5 and 20% indicates most probable cause of myocardial infarction or myocardial damage due to cardiac catheterization.

Analysis of Cerebrospinal Fluid for Proteins and Sugars

34.1 Theory

Cerebrospinal fluid (CSF) is a clear and colorless fluid that occurs between the layers pia mater and arachnoid mater covering central nervous system. CSF is formed by secretory activities of the choroid plexus, the vascular structure lying within the ventricles of brain. CSF primarily acts as a water shock absorber and also acts as a carrier of nutrients and waste products between the blood and the central nervous system. The volume of CSF formed is about 100–250 ml in adults in 24 h. The composition of CSF is same to that of brain extracellular fluid. The ionic composition for CSF is similar to the plasma for some components but differ for many other substances. Generally the composition of sodium, chloride, and magnesium in CSF is same or greater than serum, but potassium, calcium, and glucose are lower than serum. CSF glucose concentration is 60% of serum.

Note After drawing the CSF sample, the analysis of glucose and proteins should be carried out immediately. Otherwise, the specimen must be stored at $-20 °C$ which is stable for 3 days. Glucose will be rapidly destroyed in the absence of preservatives.

34.2 Analysis of Proteins in CSF

34.2.1 Pyrogallol Dye-Binding Method

34.2.1.1 Principle
Protein molecules present in CSF bind quantitatively with pyrogallol red-molybdate complex at pH 2.0 to form a violet-colored complex, which is measured at 600 nm.

34.2.1.2 Reagents
1. **Pyrogallol red dye**: To prepare the dye, dissolve 10 mg disodium molybdate, 134 mg sodium oxalate, 5.9 g succinic acid, and 430 mg sodium benzoate in

about 850 ml of distilled water. Then add 25 mg of pyrogallol red dye, and mix well to dissolve it completely. Make the final volume to 1 litre with distilled water. Store in an amber bottle.
2. **Working standard**: Prepare 1 g/dl BSA stock. Then prepare dilutions of BSA with 20, 40, 60, 80, and 100 mg/dl in normal saline.

34.2.1.3 Procedure
Take a set of seven test tubes, and label as blank, standard (S_1–S_5), and test for CSF (Ts). Add 1 ml of standard with concentrations 20, 40, 60, 80, and 100 mg/dl in tubes labelled as S_1 to S_5. To blank tube, add 1 ml of normal saline. In test (Ts), add 1 ml of CSF. Then add 3 ml of pyrogallol red in each tube. Mix well and incubate at 25°C for 15 min. Read absorbance at 600 nm.

34.2.1.4 Calculations
Plot a graph between amount of protein at x-axis and absorbance at y-axis. Extrapolate protein concentration of given CSF sample from the graph. The protein concentration can also be calculated by using equation

$$\text{Amount of protein in CSF}\,(\text{mg}/\text{dl}) = \frac{\text{OD of test} \times \text{amount of standard} \times 100}{\text{OD of standard} \times \text{volume of sample}\,(\text{ml})}$$

34.2.2 Turbidimetry Method

34.2.2.1 Principle
Proteins present in CSF are precipitating with sulfosalicylic acid, and protein amount is measured by comparing the absorbance of the turbidity at 640 nm with that of protein standards.

34.2.2.2 Reagents
1. **Normal saline**: Prepare 0.9% (w/v) NaCl.
2. **Sulfosalicylic acid** *(3 g/dl)*: Dissolve 3 g sulfosalicylic acid in a final volume of 100 ml distilled water. Reagent is stable for 6 months if stored at 25–35°C in an amber-colored bottle.
3. **Stock standard**: Prepare 1 g/dl BSA.
4. **Working standard**: Prepare dilutions of BSA with 20, 40, 60, 80, and 100 mg/dl in normal saline.

34.2.2.3 Procedure
Take a set of seven test tubes, and label as blank, standard (S_1–S_5), and test for CSF (Ts). Add 1 ml of standard with concentrations 20, 40, 60, 80, and 100 mg/dl in tubes labelled as S_1 to S_5. To blank tube, add 1 ml of normal saline. In test, add 1 ml of CSF. Then add 4 ml of sulfosalicylic acid in each tube. Mix and incubate the

contents for 5 min at room temperature. Read absorbance at 640 nm. Calculate protein amount in CSF sample as given for pyrogallol red dye method.

34.3 Analysis of CSF Glucose

Glucose determination in CSF is carried out similar to blood glucose estimation.

34.4 Clinical Significance

Normal protein levels in CSF are in range of 15–60 mg/dl. In normal conditions, albumin is main protein present in CSF, but it may contain globulins also in many diseases. The increased permeability of capillary endothelial barrier due to bacterial, viral, or fungal meningitis, multiple sclerosis, and cerebral infarction will increase CSF proteins. An increase in total CSF proteins is due to breakdown of blood-brain barrier usually as a consequence of an inflammatory reaction or obstruction in flow of CSF. In acute meningitis, polyneuritis, and tumors such as acoustic neuroma, the protein levels in CSF increase up to 0.4 g/dl. CSF/serum albumin index reveals the intactness and impairment of blood-brain barrier.

$$\text{CSF/serum albumin index} = \frac{\text{CSF albumin in mg/dl}}{\text{Serum albumin in g/dl}}$$

An index value of <9 is considered with an intact barrier. Value of 9–14 is interpreted as slight impairment while index of 14–30 as moderate impairment. An index of above 30–100 indicates severe impairment, while values greater than 100 show complete breakdown of barrier.

Normal CSF glucose range is 45–80 mg/dl. Disease conditions like tuberculosis, benign lymphocytic chronic meningitis, hypoglycemia, and metastatic tumors of meningitis cause decrease in CSF glucose levels. In bacterial meningitis, the glucose levels may disappear completely, and in tuberculosis meningitis it is usually between 10 and 40 mg/dl. Increased CSF glucose is observed in diabetes mellitus, encephalitis, brain tumors, and cerebral abscess.

To Measure Lipid Profile in Serum Sample   35

Lipids are organic substances relatively insoluble in water but soluble in organic solvents and which are related to fatty acids. They are conc. storage form of energy as well as structural components of biological membranes. The main lipids present in plasma are cholesterol, triglycerides, phospholipids, and non-esterified fatty acids. Lipids are transported in plasma and other body compartments in the form of lipoproteins. The protein part of lipoprotein is called as apolipoproteins. Each lipoprotein has a specific and relative constant composition of apolipoprotein. The apolipoprotein plays an important role in lipid transport by activating or inhibiting enzymes involved in lipid metabolism and recognizing cell membrane surface receptors. The major lipoproteins are chylomicrons, very low-density lipoprotein (VLDL), low-density lipoprotein (LDL), and high-density lipoprotein (HDL). A lipid profile typically includes the estimation of total cholesterol, HDL, LDL, VLDL, and triglycerides.

35.1 Sample Requirement

Serum or plasma sample (fasting) can be used for lipid profile analysis. A random sample can also be used if cholesterol alone has to be analyzed.

35.2 Total Cholesterol Estimation

35.2.1 Principle

Cholesterol esterase enzyme hydrolyzes cholesterol ester to free cholesterol and free fatty acids. In the presence of oxygen, free cholesterol is oxidized by cholesterol oxidase to form cholesten-4-ene-3-one and H_2O_2. The peroxidase enzyme splits H_2O_2 to water and oxygen; the oxygen thus formed oxidize phenol in presence of

4-aminophenazone to form red quinone imine dye. The color produced is measured at 520 nm and compared with that of cholesterol standard.

35.2.2 Reagents

1. **Cholesterol reagent composition**

Phosphate buffer (100 mM, pH 6.5)
4-aminophenazone (0.25 mM)
Phenol (5 mM)
Peroxidase (>5 KU/L)
Cholesterol esterase (>150 U/L)
Cholesterol oxidase (>100 U/L)
Sodium azide (0.05%)

2. **Cholesterol standard** – 200 mg/dl

35.2.3 Procedure

Take three test tubes, and label as blank, standard (S), and test (T). Pipette 1 ml cholesterol reagent to all tubes, and add 20 µl serum sample in the tube T. Then add 20 µl of 200 mg % cholesterol standard into the tube labelled as "S." Mix and incubate tubes at 37°C for 5 min and measure absorbance at 530 nm against blank. Calculate cholesterol amount in given sample.

35.2.4 Calculations

Calculate total cholesterol amount in serum sample by using equation

$$\text{Total cholesterol (mg/dl serum)} = \frac{\text{OD of test} \times \text{amount of standard} \times 100}{\text{OD of standard} \times \text{volume of sample (ml)}}$$

35.2.5 Precautions

1. Do not use lipemic or grossly lysed serum samples for the investigation.
2. Fasting blood sample is preferable.
3. If cholesterol value is more than 750 mg/dl, dilute the sample, and repeat the test.
4. Store the reagents at 2–8°C.

35.3 Triglycerides Estimation

Triglycerides are esters of glycerol possessing usually three different fatty acids. The widely distributed fats and oils in plants and animals are chemically triglycerides, and they contribute about 95% of adipose tissue. Triglycerides are transported in plasma, mostly in the form of chylomicrons and VLDL.

35.3.1 Principle

Lipases split triglycerides into glycerol and fatty acids. Glycerol kinase hydrolyzes glycerol in the presence of ATP to form glycerol-3-phosphate and ADP. Glycerol-3-phosphate oxidase enzyme splits glycerol-3-phosphate in the presence of O_2 to dihydroacetone phosphate and H_2O_2. Peroxide enzyme splits H_2O_2 to water and oxygen. The oxygen thus formed will oxidize 4-chlorophenol in the presence of 4-aminoantipyrine to form red quinone imine, which is measured at 520 nm and compared with that of triglyceride standard.

35.3.2 Reagents

1. **Buffer solution**

PIPES buffer (50 mM, pH 7.5)
4-chlorophenol (5 mM)
Magnesium ions (4.7 mM)
ATP (1 mM)
Lipases ($\geq$1 U/ml)
Peroxidase ($\geq$0.5 U/ml)
Glycerol kinase ($\geq$0.4 U/ml)
Sodium azide (0.05%)

2. **Enzyme reagent**

4-amineantipyrine (0.4 mM)
Glycerol-3-phosphate oxidase ($\geq$1.5 U/ml)
Sodium azide (0.095%)

calculated on the basis of difference between total cholesterol levels of serum and the cholesterol levels in the supernatant after centrifugation.

35.5.2 Reagents

1. **LDL-cholesterol precipitant**: Stable for 4 weeks at 20–25°C
2. **Cholesterol reagent**: Same as that used in cholesterol estimation

35.5.3 Procedure

Mix 200 µl of serum with 100 µl of precipitating reagent and allow standing for 15 min at room temperature and then centrifuge at 10000 rpm for 2 min or for 15 min at 1500 rpm.

Take two test tubes, and mark them as test and blank. Add 25 µl of supernatant in tube labelled as test and 25 µl of distilled water in blank tube. Add 1 ml of cholesterol reagent to these tubes, mix, and incubate at 37°C for 5 min or room temperature for 10 min. Measure absorbance at 500 nm against blank in a spectrophotometer, and calculate the cholesterol value in the supernatant. From total cholesterol value, the value obtained in the supernatant is subtracted to get the LDL cholesterol value.

$$LDL = HDL + \frac{Triglycerides}{5}$$

Here triglyceride/5 gives VLDL value.

35.6 Clinical Significance

35.6.1 Cholesterol Levels

A cholesterol value below 200 mg/dl is always considered desirable and indicates a low risk of heart disease. Cholesterol value above 240 mg/dl is considered high risk. Cholesterol is high during pregnancy and up to 6 weeks after the baby is born. Some drugs like anabolic steroids, epinephrine, oral contraceptives, and vitamin D are known to increase cholesterol levels.

35.6.2 LDL Values

LDL cholesterol below 100 mg/dl is desirable. LDL greater than 150 mg/dl indicates increased risk to cardiovascular diseases. Out of all the forms of cholesterol in the blood, LDL levels are considered the most important factor in determining risk of heart diseases.

35.6.3 HDL Values

Normal range of HDL cholesterol is 30–60 mg/dl in the human blood. HDL less than 40 mg/dl indicates high risk of heart disease. HDL is called good cholesterol since it transports cholesterol from peripheral tissues for its degradation, and hence it removes excess cholesterol from tissues.

35.6.4 Triglycerides

A normal level of triglycerides is less than 150 mg/dl in fasting condition. It is unusual to have high triglycerides without having high cholesterol. High levels of triglycerides increase the risk of developing pancreatitis.

36 To Estimate Sodium and Potassium in Serum by Using Flame Photometer

36.1 Theory

Sodium is the major cation of extracellular fluid. Sodium is important for the maintenance of osmotic pressure, fluid balance, and acid-base regulation. The common salts used in cooking and food intake are the major sources of sodium. The sodium is lost from the body via excretion through kidneys and excessive sweating. About 50% of body sodium is present in bones followed by 40% in extracellular fluid and 10% in soft tissues. Both extracellular and intracellular sodium are the exchangeable form of sodium. The intracellular sodium contributes about 10 mmol/L, and the extracellular form contributes about 135–145 mmol/L. Sodium is also involved in normal muscle irritability, cell permeability, and maintenance of heartbeat. In contrast to sodium, potassium is present mainly in intracellular fluid. It is present majorly in RBCs (23 times higher than plasma). Like sodium, excess potassium is also excreted by kidneys. Normal potassium concentration in plasma is in range of 3.5–5.5 mEq/L. It is necessary for regulation of intracellular osmotic pressure, water balance in cells, transmission of nerve impulse, etc.

36.2 Specimen Type, Collection, and Storage

Serum sample is used. Sodium and potassium present in serum remain stable for several hours at 25–35 °C and for almost 3 months if stored at −20 °C. Lithium heparin may be preferred as an anticoagulant, but anticoagulants containing sodium or potassium salts should be avoided. Hemolysis will increase potassium levels. Dilute the standards and serum sample 1:100 using distilled water before use.

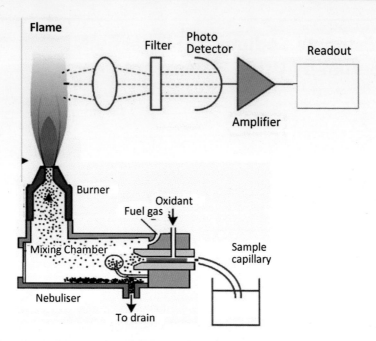

Fig. 36.1 Schematic representation of flame photometer

36.3 Principle

"The test solution is passed as fine spray into the flame. The flame evaporates the sample and dissociates the salts to give neutral ions. The heat energy of the flame excites some of these atoms which cause electrons to move to a higher energy level. The electrons are unstable in the excited state and hence return to lower energy level. In this process, the electrons will emit a light of a fixed wavelength of electromagnetic spectrum. The amount of light emitted is directly proportional to the number of atoms undergoing in excited state. The number of electrons in excited state in turn is directly related to the concentration of the substance in the sample. Sodium emits yellow light of 589 nm, whereas potassium emits violet light of wavelength maximum 404.4 nm and 765.9 nm (Fig. 36.1).

36.4 Reagents

1. **Stock sodium 1000 mmol/L:** Dissolve 5.85 g of NaCl in final volume of 100 ml distilled water.
2. **Stock potassium 100 mmol/L:** Dissolve 746 mg of dried KCl in final volume of 100 ml distilled water.

3. **Working standards:** Mixed working standards for sodium and potassium are prepared as follows:
   (a) **Sodium/potassium (110/2 mEq/L):** This standard is prepared by mixing 11 ml of stock sodium and 2 ml of stock potassium standards in 87 ml distilled water. The final 100 ml mixture contains 110 mEq of sodium and 2 mEq of potassium.
   (b) **Sodium/potassium (140/4 mEq/L):** To prepare the standard, take 14 ml of stock sodium and 4 ml of stock potassium standards in a flask, and add 82 ml distilled water to make final volume to 100 ml.
   (c) **Sodium/potassium (170/8 mEq/L):** Take 17 ml of stock sodium and 8 ml of stock potassium standards in a flask, and add 75 ml distilled water to prepare standard.

36.5 Procedure

1. The flame photometer is switched on, and appropriate filter is selected with the help of filter selector wheel.
2. Put the air compressor on, and adjust air pressure between 0.4 and 0.6 kg/cm^2.
3. Introduce glass-distilled water through atomizer.
4. Then open the gas cylinder; the flame is ignited after removing the trapper at the rear of the flame photometer. Adjust it so that flame is divided into fine sharp cones.
5. Make zero adjustment by introducing distilled water.
6. Aspirate mixed Na$^+$ and K$^+$ standard (110/2 mEq/L), and adjust knob meant for sodium to digits 110 and by knob meant for potassium to digit 2.
7. Introduce standard 140/4 mEq/L. The digital display will show exact reading if standards are correct.
8. Then aspirate third standard 170/8 mEq/L for sodium and potassium, and adjust reading.
9. Now introduce the diluted test serum sample, and record readings for sodium and potassium.

36.6 Clinical Significance

Normal sodium levels in serum are 135–145 mmol/L, while serum potassium level is 3.5–5.0 mmol/L. Increase in sodium concentration more than 145 mmol/L is called hypernatremia. Hypernatremia is caused by hyperactivity of adrenal cortex, loss of water due to dehydration, high administration of sodium salts, and steroid therapy. The decrease in serum sodium levels (<135 mmol/L) is called hyponatremia. It may occur due to prolonged vomiting, diarrhea, severe polyuria, diuretic medication, and metabolic acidosis.

When the plasma potassium concentration exceeds 5.5 mmol/L, the condition is called hyperkalemia. Higher potassium levels cause mental confusion, weakness, numbness, slowed heart rates, and vascular collapse. Very high potassium levels are fatal. Hyperkalemia occurs in dehydration, diabetic ketoacidosis, massive intravascular hemolysis, and violent muscular activity. The plasma potassium levels less than 3.5 mmol/L lead to hypokalemia. Hypokalemia is caused by prolonged diarrhea, vomiting, and also in renal tubular acidosis. The symptoms include muscle weakness, irritability, paralysis, and cardiac arrest.

To Perform Radioimmunoassay 37

Radioimmunoassay (RIA) is a highly sensitive and specific analytical tool used for detection of antigen or antibody in a sample. It was developed in the late 1950s by Rosalyn Yalow and Solomon A. Berson for estimation of insulin in human serum.

37.1 Principle

Radioimmunoassay combines the principles of radioactivity of isotopes and immunological reactions of antigen and antibody. The principle is based on the competition between labelled and unlabelled antigens to bind with antibody to form antigen-antibody complexes. The test is employed to determine the concentration of unlabelled antigens. The antibody specific for antigen is incubated with unlabelled antigen in the presence of high concentration of antigens labelled with isotope (say I^{131}). The presence of both labelled and unlabelled antigens in the sample mixture creates a competition between these antigens to bind with antibody. Initially labelled antigens will bind in more amounts due to their high concentration present in sample. Then amount of unlabelled antigen is increased progressively in the sample. As the concentration of unlabelled antigens increases, they will replace the labelled antigens from their already existing antigen-antibody complex. Thus, the concentration of labelled antigens bound with antibody will be inversely related to unlabelled antigens, i.e., the higher is the concentration of labeled antigens bound with antibody, the less is the concentration of unlabelled antigens or vice versa. The concentration of labelled antigen can be determined by precipitating them and determining radioactivity present in this antigen-antibody complex. A standard curve is drawn using different concentrations of unlabelled antigens and same quantities of antibody and labelled antigens. From this plot the amount of unlabelled antigens present in the test sample may be determined precisely (Fig. 37.1).

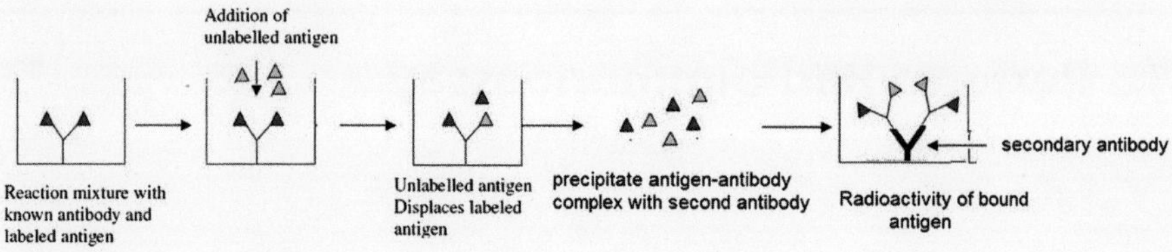

Fig. 37.1 Principle of RIA

37.2 Advantage

Advantage of RIA is its high sensitivity. It can measure up to few pictograms of antigens. The specificity of the assay depends upon the specificity of antiserum.

37.3 Applications

Radioimmunoassay is used to estimate hormones and proteins possessing antigenic properties. These days, applications of RIA have been extended for detection of peptides, steroid hormones, drugs, neurotransmitters, antibodies, and structural proteins. It is also important for blood bank screening for the hepatitis virus, early cancer detection, measurement of growth hormone levels, and diagnosis and treatment of peptic ulcers.

To Perform Enzyme-Linked Immunosorbent Assay

38

Enzyme-linked immunosorbent assay (ELISA) is an immunoassay that combines the specificity of antibody and antigen usually detected in presence of enzyme conjugated with antibody. The enzyme linked with the antibody reacts with specific substrate to produce colored product which is related to antigen or antibody concentration.

38.1 Principle

Enzyme-linked immunosorbent assay is based on immunochemical principles of antigen-antibody reaction. The antibody to be determined is coated on an inert surface. The sample containing protein is applied on antibody-coated surface. The protein-antibody complex is then reacted with second antibody specific to protein. This second antibody is enzyme labelled (alkaline phosphatase, horseradish peroxidase, or β-galactosidase). The unbound enzyme-linked second antibody is washed, and enzyme activity is determined by adding suitable substrate specific to enzyme bound with second antibody (i.e., diaminobenzidine for horseradish peroxidase). This gives a colored product that is directly proportional to the protein being estimated (Fig. 38.1).

The above mentioned method is called indirect ELISA. Other variants of ELISA are also in use.

38.2 Sandwich ELISA

Sandwich ELISA is named so because the analyte to be measured is bound between two primary antibodies – the capture antibody and the detection antibody. In this process, the antibody is immobilized on microwell plate, and the sample containing antigen is added to this well that reacts with immobilized antibody. The well is washed to remove unbound antigen, and second antibody that has been conjugated

Fig. 38.1 Principle of ELISA

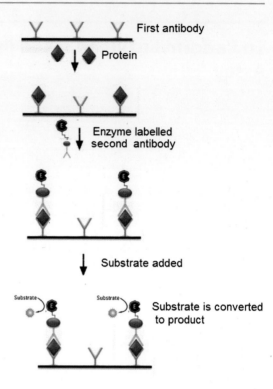

with enzyme is added. The substrate specific to enzyme is added to well, and colored product formed is measured. The sandwich ELISA is useful to detect antigen concentration.

38.3 Competitive ELISA

Competitive ELISA is also used to measure antigen concentration. The process involves the incubation of antibody in a sample solution containing antigen. This antigen-antibody mixture is then transferred to a microplate well coated with antigen. The assay is based on the fact that the presence of more number of antigens in the sample will bind with more number of antibodies, and hence available free antibody concentration that can bind with the antigen-coated well will be less. The concentration of primary antibody that binds with antigen in the well can be detected by addition of enzyme-conjugated secondary antibody specific for the isotype of the primary antibody. The absorbance for this reaction will be inversely proportional to the concentration of antigen in the original sample.

38.4 Direct ELISA

In direct ELISA, the antigen is immobilized on the microwell plate. Then an enzyme-labelled primary antibody is added to the well that reacts directly with antigen. Direct detection ELISA is not a frequently used method but is used commonly for immunohistochemical staining of tissues and cells.

38.5 Applications of ELISA

Enzyme-linked immunosorbent assay is an important tool to find the presence of antigen or antibody in a sample. It is used commonly to assay various proteins including insulin, luteinizing hormone, etc. Indirect ELISA is used for diagnosis of human immunodeficiency virus. It can also be used for the screening of monoclonal antibodies. The most commonly used pregnancy test is based on ELISA which detects human chorionic gonadotropin in urine. It is also being used in food industry for detection of potential food allergens.

Some Important Case Studies

39.1 Case Studies of Sugar Impairment

Q.1. A diabetic patient on insulin therapy was brought to the hospital in a semiconscious state. He was sweating profusely and had tremors of hands. What is the possible diagnosis?

Ans: The possibility is hypoglycemia due to excessive dose of insulin.

Q.2. An apparently healthy man underwent for routine laboratory investigations. The results showed that his fasting blood sugar level was 85 mg/dl while urine constituents were normal. However, his postprandial blood sugar after one-and-half hours of food intake was observed to be 155 mg/dl, and urine sample at that time was also positive for Benedict's test. What is the diagnosis?

Ans: The case symptoms indicate toward renal glycosuria. In renal glycosuria, glucose is excreted in the urine despite the fact that blood glucose levels are normal or even low. With normal kidney functions, only very high blood glucose levels can cause its excretion in urine. So, renal glycosuria suggests abnormal functioning of the renal tubules. Normal renal threshold for glucose is 180 mg/dl, but in this case renal threshold is lowered.

Q.3. A 35-year-old female was examined for the laboratory tests. The report showed that her blood sugar (fasting) was 90 mg/dl and blood sugar (P.P.) was 168 mg/dl. The urinary analysis showed absence of urine sugar (fasting), but in postprandial state urine was significantly positive for sugar. Comment on report.

Ans: In this case, the fasting blood sugar is within normal limits, and urine is negative for reducing sugars. However, under postprandial condition, the urine tests positive for sugar although blood sugar is less than the normal renal threshold for glucose (180 mg/dl). This indicates that renal threshold for glucose in this patient has been lowered and it is a case of renal glycosuria.

Q.4. A 45-year-old man was hospitalized with complain of dizziness and weakness. His random blood sugar was 40 mg. History revealed that he had the skipped breakfast. What is the probable diagnosis?

Ans: Patient has hypoglycemia, probably due to fasting. Under physiological conditions, brain derives energy from glucose. Hypoglycemia is considered when blood glucose levels fall below 60 mg/dl, and symptoms begin at this concentration of glucose. The brain symptoms appear when glucose level falls below 50 mg/dl.

39.2 Case Studies of Diabetic Ketoacidosis

Q 1. A 45-year-old male was admitted to hospital in an unconscious state. He appeared moderately dehydrated with increased pulse rate and acetonemic breath. The urine analysis report showed presence of high amount of sugar, and Rothera's test was positive for ketone bodies. What is the probable diagnosis?

Ans: Excretion of sugar in urine is accompanied by polyuria, which therefore causes dehydration. This is a case of diabetic ketoacidosis. High levels of ketone bodies in blood cause acetonemic breath and unconsciousness.

Q.2. A patient was admitted in the hospital in unconscious state. His blood glucose level was 290 mg/dl. His urine was positive for both Rothera's test and Benedict's test.
What is the diagnosis?

Ans: High blood glucose levels and presence of ketone bodies in urine indicate diabetic ketoacidosis which is a life-threatening complication arising due to diabetes mellitus. Ketone bodies in urine occur due to insulin deficiency and less utilization of glucose by cells.

Q.3. A young boy was admitted to hospital in a delirious state. The examination showed that he was breathing rapidly and appeared dehydrated. Laboratory investigation results showed high blood sugar levels (325 mg/dl) and high serum K^+ (6.5 mEq/L) levels, but serum sodium was normal. Urine test was positive both for ketone bodies and reducing sugars. Give comment on diagnosis.

Ans: Increased blood glucose value with presence of sugar and ketone bodies in urine is indicative of diabetic mellitus. Hyperkalemia is seen in insulin deficiency as K^+ uptake by cells requires insulin.

Q.4. A known diabetic patient was admitted to hospital in a semiconscious state. He had cellulitis on the right foot. The laboratory investigation on blood sample showed:

Glucose: 375 mg/dl
Urea: 85 mg/dl
Creatinine: 2.1 mg/dl
Sodium: 135 mmol/L
Potassium: 5.6 mmol/L
Chloride: 95 mmol/L
Urine was positive for glucose and ketone bodies. Discuss the results.

Ans: This is a case of uncontrolled diabetes mellitus with ketoacidosis. Diabetic patients with chronic hyperglycemia may have prominent skin and soft tissue infections. High serum glucose levels may lead to hyponatremia. Ketosis and acidosis elevate the serum potassium levels by causing a shift of potassium from the intracellular to the extracellular fluid. Urea and creatinine are increased due to prerenal acute kidney injury.

Q.5. A 45-year-old woman with symptoms of loss of weight, polyphagia, polydipsia, and polyuria was brought to the hospital. Her laboratory reports depicted:

Serum glucose (Fasting) – 165 mg/dl
Serum glucose (P.P.) – 225 mg/dl
The urine analysis showed absence of sugar (fasting), but sugar was present significantly in postprandial stage. Comment on report.

Ans: Fasting and postprandial blood sugar levels are increased above normal range, while postprandial urine sugar is positive. This indicates diabetes mellitus. Loss of weight, polyphagia, polydipsia, and polyuria are classical symptoms of diabetes.

39.3 Case Studies of Calcium and Phosphate Impairments

Q.1. The laboratory investigation report of a 55-year-old woman's blood sample revealed high calcium (12.8 mg/dl) and low potassium (2.2 mg/dl) levels. The alkaline phosphatase activity was 45 KAU. What is the probable diagnosis?

Ans: Increased calcium and alkaline phosphatase and decrease in phosphorus are suggestive of hyperparathyroidism.

Q.2. A child was brought to pediatric OPD. He had bowlegs, knock-knees, and protruding abdomen and had suffered mild convulsions. The serum calcium level was very low. What is the probable diagnosis?

Ans: Low serum calcium levels lead to soft bones which tend to bend and delay in walking. Hypocalcemia also causes convulsions. This is a probable case of rickets.

39.4 Case Studies of Protein Energy Malnutrition

Q.1. A 3-year-old child was observed to have mild generalized edema, intermittent diarrhea, rough skin, and coarse hair and weakness. Laboratory investigation data showed:

Hemoglobin – 8 mg%
Total proteins (serum) – 4.9 g/dl
Albumin (serum) – 1.7 g/dl
Urine constituents – normal
What is the probable diagnosis?

Ans: Generalized edema, intermittent diarrhea, rough skin, and coarse hair indicate protein-deficient diet intake. This appears to be a case of protein malnutrition called Kwashiorkor. Kwashiorkor is characterized by very low serum albumin levels that lead to edema. Low hemoglobin levels indicate anemia due to dietary deficiency of iron.

Q.2. A 2.5-year-old child with history of poor eating for the last 1 month and having intermittent diarrhea was brought to the hospital. The physical examination showed that he was underweight with retarded growth and his abdomen was distended, liver was moderately enlarged, and generalized edema was present. The blood analysis showed:

Hemoglobin – 6.2 gm/dl
Total protein – 5.2 gm/dl
Albumin – 1.6 gm/dl
What is the probable diagnosis?

Ans: The child is suffering from protein energy malnutrition (kwashiorkor). The present case indicates a severe form of malnutrition. The presence of edema, irritability, dermatitis, enlarged liver with fatty infiltration, inadequate growth, muscles loss, vomiting, and diarrhea are classical symptoms of kwashiorkor.

39.5 Case Studies of Gout/Uric Acid

Q.1. A 45-year-old, alcoholic patient was complaining of severe pain in joints and toes. His laboratory report for blood analysis presented:

Sugar (random) – 88 mg/dl
Uric acid – 9.8 mg/dl
Urea – 40 mg/dl

39.5 Case Studies of Gout/Uric Acid

Creatinine – 1.8 mg/dl
Comment on report.

Ans: The presented reports show increased uric acid levels in serum. High uric acid levels indicate gout. In gout, uric acid is deposited in joints leading to inflammation and pain. Here, increased uric acid in serum does not correlate with normal creatinine; hence possibility of kidney dysfunction is ruled out.

Q.2. A 1-year-old baby having retarded development and showing habits of lips and finger biting was brought to hospital. The laboratory investigation of blood sample revealed very high uric acid levels (9.5 mg/dl), while hemoglobin, urea, creatinine, and sugar were within normal range. Comment on the report.

Ans: The clinical features of delayed development and high serum uric acid levels are suggestive of Lesch-Nyhan syndrome. In this syndrome, the deficiency of the enzyme hypoxanthine-guanine phosphoribosyltransferase causes high uric acid levels in body fluids. The combination of increased synthesis and decreased utilization of purines leads to high uric acid levels, a condition called hyperuricemia.

Q.3. A 45-year-old man was complaining of severe pain in the right toe. He did not reveal any history of pain in any joints previously. He had been binge drinking the previous night. On examination, he had fever and was in distress due to the pain. The right toe was swollen, warm, red, and very tender. Synovial fluid analysis revealed needle-shaped crystals under microscopy. Serum uric acid level was 8.9 mg/dl, and 24 hr urinary uric acid excretion was 500 mg/dl. Random blood sugar was 140 mg/dl. Other tests were normal. What is the likely diagnosis?

Ans: The likely diagnosis is gouty arthritis. The pain in the big toe is precipitated by alcohol intake, which is a typical feature of gouty arthritis. Serum uric acid and synovial fluid analysis results are confirmatory. Gout is characterized by hyperuricemia. Monosodium urate crystals are deposited in joints and connective tissues, and there is risk of uric acid nephrolithiasis. Acute gouty arthritis can be triggered by trauma, stress, vascular occlusions, surgery, drugs, and purine-rich food including alcohol. Hyperuricemia can be due to increased production or reduced excretion of uric acid or a combination of these two.

Q.4. A 40-year-old male consumed much food and alcohol during dinner. The next morning, he felt excruciating pain in the ankle. On examination, he had fever. His ankle joint was red and swollen and was very tender and stiff. No other joints were involved. Lymph glands were normal. The laboratory data was:

Blood glucose: 120 mg/dl
Blood urea: 40 mg/dl
Serum creatinine: 1.2 mg/dl
Serum uric acid: 10.5 mg/dl

Urine pH: 6.0
What is the most probable diagnosis?

Ans: The likely diagnosis is gouty arthritis. Gout classically afflicts the big toe, but other joints may also sometimes be affected.

39.6 Case Studies of Liver Functions

Q.1. A 48-year-old, fat female was admitted to hospital with complain of abdominal pain, vomiting, nausea, and loss of appetite. She passed dark yellow urine and clay-colored stool. The laboratory blood examination revealed:

Total bilirubin – 12 mg/dl
Indirect (unconjugated) bilirubin – 2.2 mg/dl
Direct (conjugated) bilirubin – 9.8 mg/dl
ALT – 105 U/L
AST – 80 U/L
ALP – 160 U/L

Urine was positive for bile salts and bile pigments. Urobilinogen was absent in urine. Comment on report.

Ans: The data shows that total bilirubin and conjugated bilirubin are very high as compared to unconjugated bilirubin. Serum ALT and AST levels are highly increased (ALT levels are higher than AST). This is a case of obstructive jaundice. In this case, ALP is also increased significantly. Increase in serum ALP activity is usually related to hepatobiliary obstruction. In hepatobiliary disease, the concentration of ALP is increased due to defect in excretion of ALP through bile. ALP mainly produced in the bone tissue reaches to the liver and excreted through bile. The presence of conjugated bilirubin in the urine and excretion of urobilinogen-free urine suggest the case of obstructive jaundice. When complete obstruction of the bile duct occurs, urobilinogen is absent in urine because bilirubin is not transported to the intestine. Normally, bilirubin is converted to urobilinogen by bacterial action in the intestine, and a portion of it is absorbed by enterohepatic circulation. Urobilinogen is also converted to urobilin in the kidneys and excreted. The pale stools and dark urine also confirm obstructive or posthepatic cause since feces color is contributed from bile pigments.

Q.2. A 45-year-old fat woman complained of recurrent pain in the abdomen which often aggravated by fatty food intake. She had very high blood bilirubin levels. The urine examination showed the presence of bile pigments and bile salts, but urobilinogen was absent. What is the most likely cause?

39.6 Case Studies of Liver Functions

Ans: This is a case of obstructive jaundice due to cholelithiasis. Obstructive jaundice is characterized by the absence of urobilinogen in urine as explained above. In present case, the obstruction of bile ducts by gallstones may have caused the absence of urobilinogen in urine.

Q.3. A 55-year-old man diagnosed with jaundice was brought to hospital. He had complain of weight loss and passed pale stools. The laboratory investigation showed:

Bilirubin – 18 mg/dl
ALT – 95 U/L
AST – 75 U/L
ALP – 330 U/L
Serum glucose – 78 mg/dl
Urine was positive for bile salts and bile pigments. Give the probable diagnosis.

Ans: Here bilirubin levels are very high with moderate elevation of AST and ALT and greater increase of ALP. The urine shows the presence of bile salts and bile pigments. All these diagnosis indicate obstructive jaundice.

Q.4. A young boy was brought to the hospital with history of vomiting, nausea, loss of appetite, and abdominal pain. The laboratory data showed:

Bilirubin total – 7.5 mg/dl
Indirect bilirubin – 6.5 mg/dl
Direct bilirubin – 4.2 mg/dl
ALT – 78 U/L
AST – 88 U/L
ALP – 110 U/L
Urine was positive for bile pigments and urobilinogen and very dark in color. Comment on report.

Ans: Bilirubin, AST, ALT, and ALP levels are significantly increased. Urine is positive for bile pigments and urobilinogen. So, present case indicates hepatocellular or hepatic jaundice. In hepatic jaundice, levels of both conjugated and unconjugated bilirubin rise in the blood. The abnormally high amount of conjugated bilirubin and bile salts leads to their excretion in the urine. This conjugated bilirubin provides the urine the dark color.

Q.5. A 35-year-old man with diagnosed jaundice and history of drug and alcohol abuse was brought to hospital. He showed pitting edema. The blood investigation report revealed:

Total protein – 8.5 g/dl
Albumin – 2.0 g/dl
Globulins – 5.8 g/dl
Bilirubin – 6.5 mg/dl

ALT – 305 U/L
AST – 210 U/L
ALP – 175 U/L
Blood urea nitrogen – 6.0 mg/dl
Urine was positive for bilirubin. Comment on report.

Ans: The laboratory data presents with abnormal liver function tests. The significant elevation of AST and ALT reflects chronic liver disease. In acute liver disease, transaminase levels are much higher but remain moderately higher in chronic liver disease due to destruction of hepatocytes. Low value of albumin and blood urea nitrogen suggests impaired synthesis due to loss of hepatocytes. Overall, it is a probable case of chronic hepatitis.

Q.6. A case with the following laboratory results was presented in a meeting:

Serum bilirubin – 9.0 mg%
Conjugated bilirubin – 0.5 mg%
Unconjugated bilirubin – 8.5 mg%
ALT – 30 U/L
AST – 35 U/L
ALP – 10 KAU
Urine showed negative test for bile salts and bile pigments but positive (++) for urobilinogen. Feces stercobilinogen was also positive. Comment on the case, and give the provisional diagnosis.

Ans: This is a case of hemolytic jaundice caused by high destruction of red blood cells. In prehepatic or hemolytic jaundice, the unconjugated bilirubin is increased due to increased destruction of red blood cells, and the liver is unable to cope with the increased demand for conjugation. Hence, deposition of this unconjugated bilirubin occurs into various tissues leading to a jaundiced appearance. Urine is deprived of bilirubin because unconjugated bilirubin is not water-insoluble, so increased urine urobilinogen without bilirubin in urine is suggestive of hemolytic jaundice. The increased hemolysis of blood cells produces high bilirubin leading to the increased excretion of urobilinogen in urine.

Q.7. A 30-year-old man was diagnosed with the presence of gallstones. He had severe abdominal pain also. The laboratory investigations showed:

RBCs: 3 million/cu mm
Reticulocytes: 14%
Hemoglobin: 8.2 g/dl
Serum bilirubin (total): 2.6 mg/dl
The urine was positive for urobilinogen. Comment on probable diagnosis.

Ans: The case indicates severe hemolytic anemia, leading to bilirubinemia. Excretion of large quantities of bilirubin through bile leads to gallstones, made up of bilirubin. Decreased life span leads to active generation of RBCs, hence increased reticulocytes.

39.7 Case Studies of Kidney Functions

Q.1. A child having history of fever and abdominal pain since few days was examined by doctors. He passed red-colored urine since 3 days and also having pedal edema. His blood examination results were:

Urea – 95 mg/dl
Creatinine – 4.4 mg/dl
Cholesterol – 230 mg/dl
Total proteins – 5.8 g/dl
Albumin – 2.2 g/dl
The urine analysis showed the presence of albumin and blood. Comment on report.

Ans: The increased serum urea and creatinine levels indicate compromised kidney functions. Serum total proteins and albumin levels are decreased, while urine has albumin and blood. This indicates acute glomerulonephritis. The damaged kidney glomeruli can result in nephritic syndrome. Nephritic syndrome symptoms include low blood albumin causing edema. Fever and abdominal pain are due to kidney dysfunction.

Q.2. An old man with history of diabetes since 20 years was admitted to hospital. His urine output was very low and blood pressure was 220/100 mmHg. The blood investigation results were:

Sugar (random) – 300 mg/dl
Total proteins – 5.0 g/dl
Urea – 200 mg/dl
Creatinine – 12 mg/dl
Cholesterol – 290 mg/dl
Na^+ – 125 mEq/L
K^+ – 5.6 mEq/L
Urine was tested positive for albumin. Comment on report.

Ans: This is a case of chronic renal failure arising from diabetic nephropathy due to chronic primary hyperglycemia. Increased urea and creatinine levels and proteinuria indicate renal damage. Proteinuria is associated with renal injury, accompanied by diabetic nephropathy. The renal damage may increase permeability of glomeruli further allowing plasma proteins to escape into the urine. Chronic renal damage also elevates blood pressure. Decrease in sodium (hyponatremia) and increased potassium (hyperkalemia) levels further confirm diagnosis.

Q.3. A 6-year-old boy with complain of fever and facial edema was brought to hospital. He appeared to be lethargic and pale. The laboratory blood results were:

Sugar (random) – 80 mg/dl
Total proteins – 5.0 g/dl
Albumin – 2.2 g/dl
Urea – 42 mg/dl
Creatinine – 1.9 mg/dl
Cholesterol – 380 mg/dl
Na^+ – 125 mEq/L
K^+ – 6.0 mEq/L

Urine was tested positive for protein and hematuria was present. Comment on report.

Ans: The laboratory results show that urea and creatinine are near normal while total protein and albumin are decreased. Urine tests positive for proteins, and there is significant increase in serum cholesterol. These findings are suggestive of nephritic syndrome that generally occurs in glomerulonephritis. In glomerulonephritis, the glomerular basement membrane becomes thin, and there is the presence of small pores in podocytes of the glomerulus. These pores allow proteins and RBCs to pass into the urine. Nephritic syndrome also decreases serum albumin levels due to its high excretion in urine.

Q.4. A 50-year-old hypertensive patient with facial and lower limb edema appearance was admitted to hospital. Laboratory blood analysis results showed:

Sugar (fasting) – 250 mg/dl
Albumin – 2.2 gm/dl
Cholesterol – 275 mg/dl
Creatinine – 2.4 mg/dl
Urea – 105 mg/dl

Urine examination showed positive Benedict's test with red precipitate, and urine proteins were very high. What is the possible diagnosis?

Ans: Proteinuria, hypoalbuminemia, and hypercholesterolemia along with acute renal failure and edema are the classical presentation of nephrotic syndrome. Patient has diabetes mellitus (urine positive for reducing sugars) and hypertension, and these may have caused nephrotic syndrome.

Q.5. The blood investigation of a 70-year-old woman is shown below:

Sodium – 125 mmol/L
Potassium – 3.8 mmol/L
Urea – 205 mg/dL
Creatinine – 4.4 mg/dL
Calcium – 6.2 mg/dL
Phosphate – 7.5 mg/dL
Alkaline phosphatase – 90 U/L

What is the most probable diagnosis?

Ans: Increase in blood urea and creatinine indicates renal failure. The commonest cause of hyperphosphatemia is renal failure. High phosphate levels in this case indicate a severe degree of renal insufficiency. High phosphorus interferes with calcium absorption, resulting in hypocalcemia and renal osteodystrophy.

39.8 Case Studies of Cardiac Functions

Q.1. A 55-year-old man was complaining of chest pain and labored breathing. The laboratory blood examination report was:

CK – 690 U/L
CK-MB – 88 U/L
AST – 90 U/L
LDH – 785 U/L
Cholesterol – 360 mg/dl
HDL cholesterol – 35 mg/dl
LDL cholesterol – 260 mg/dl
Comment on report.

Ans: In acute myocardial infarction, chest pain is the commonest symptom. Pulmonary edema is the consequence of damage to the heart limiting the output of the left ventricle and shortening of breath. Here, the serum levels of CK, CK-MB, AST, and LDH are increased which suggest myocardial infarction. The serum cholesterol and LDL cholesterol levels are very high, while HDL levels are decreased which predispose a person to heart disease.

Q.2. A 58-year-old man with known diabetic history since the last 12 years underwent laboratory examination. The blood test results showed that his AST, ALT, urea, and creatinine levels were normal while fasting sugar was 150 mg/dl. The lipid profile was:

Cholesterol – 310 mg/dl
HDL cholesterol – 32 mg/dl
LDL cholesterol – 220 mg/dl
Triglyceride – 310 mg/dl
Comment on report.

Ans: The person has uncontrolled diabetes mellitus. His liver and kidney appear to be normal as reflected by normal AST, ALT, urea, and creatinine levels. However, serum cholesterol and LDL cholesterol levels as well as triglyceride levels are high indicating hyperlipidemia. In diabetes, increased lipolysis due to low insulin and high glucagon levels leads to excessive acetyl CoA production which is diverted for cholesterol and triglyceride synthesis.

Suggested Readings

1. Tietz NW (1995) Clinical guide to laboratory tests, 3rd edn. W.B. Saunders Company, Philadelphia
2. Wallach J (2007) Interpretation of diagnostic tests, 8th edn. Wolters Kluwer/Lippincott Williams & Wilkins, Philadelphia
3. Chawla R (2014) Practical clinical biochemistry: methods and interpretations, 4th edn. Jaypee Brothers Medical Publishers (P) Ltd, New Delhi
4. David T, Plummer DT (2004) An introduction to practical biochemistry, 3rd edn. Tata McGraw Hill Education Pvt. Ltd, New Delhi
5. Basten G (2011) Introduction to clinical biochemistry: interpreting blood results. Ventus Publishing APS
6. Lehmann CA (1998) Saunders manual of clinical laboratory science, 1st edn. W.B. Saunders Company, Philadelphia
7. Ramakrishnan S (2012) Manual of medical laboratory techniques, 1st edn. Jaypee Brothers Medical Publishers (P) Ltd, New Delhi
8. Reinhold JG (1953) Manual determination of serum total protein, albumin and globulin fractions by Biuret method. In: Reiner M (ed) Standard methods in clinical chemistry. Academic Press, New York, pp 88–97
9. Mohanty B, Basu S (2006) Fundamentals of practical clinical biochemistry. B. I. Publications, Pvt. Ltd, New Delhi

Index

A
Absorbance, 17–19, 40, 41, 45–47, 59, 62, 66, 69, 77, 83, 87, 91, 95, 99, 100, 105, 109, 129–132, 136, 137, 140, 142–144, 154
Accuracy, 9, 11–13
Acetate buffer, 24, 25
Acetoacetic acid, 119–122
Acid phosphatase (ACP)
　clinical significance, 112
　distribution, 111, 112
　normal range, 112
Acylglycerol, 117
A/G ratio, 43–48
Alanine amino transferase (ALT)
　activity measurement, 104, 106
　chemical reaction, 104
　clinical significance, 106
　distribution, 103–106
Alkaline copper sulfate, 61
Alkaline phosphatase (ALP), 88, 107–109, 111, 112, 153, 159, 166
Allantoin, 82
Amino napthol sulphonic acid (ANSA), 90, 91
Amylase
　diagnostic importance, 114, 115
　normal value, 114, 115
　occurrence, 113–115
　starch hydrolysis, 114
　types, 113
Anticoagulant (s)
　EDTA, 6, 7, 107
　heparin, 147
　sodium fluoride, 7
　sodium/potassium oxalate, 7
Anuria, 30
Apolipoprotein, 139
Apoproteins, 93
Aspartate amino transferase (AST)
　activity measurement, 104, 106
　chemical reaction, 104
　clinical significance, 106
　distribution, 103
Autoanalyzer, 15
Automation, 13–15

B
Bad cholesterol, 143
Bar coding, 14
Batch analyzer, 14
Beer-Lambert law, 18
Bence-Jones proteins, 34–36
Benedict's test, 49, 157, 158, 166
β-glucuronidases, 123
β-hydroxybutyric acid, 119
Bile pigments
　clinical significance, 125
　formation, 123, 125
　test (s), 123–127
　transport, 123
Bilirubin
　chemical reaction, 98
　diagnostic importance, 127
　excretion, 39, 101, 162, 164
　reference range, 100
　synthesis, 124, 125
　transport, 97, 123
Bilirubin diglucuronide, 123
Biliverdin, 97, 123
Biuret, 32, 39–41, 44, 45
Blood
　collection, 5–7, 14, 72
　hemolysis, 5, 90, 92, 164

Blood (*cont.*)
    pH, 5, 25, 122
Bradshaw's test, 36
Brij 35, 46
Bromo cresol green (BCG), 46–48
Buffers, 21–27, 33, 34, 46, 58, 64, 65, 86, 87, 104, 105, 108, 111, 112, 118, 130, 132, 140–142

C
Calcium
    clinical significance, 87
    distribution, 86
    functions, 85
    hypercalcemia, 88
    hypocalcemia, 88, 159, 167
    ionized, 85, 88
    protein bound, 85
    reference range, 85, 159
Calibration, 9, 10, 12, 104
Capillary blood, 5
Carbonate-bicarbonate buffer, 25, 108, 112
Case studies
    calcium & phosphorus impairment, 159
    cardiac functions, 167
    diabetic ketoacidosis, 158
    gout/uric acid, 160
    kidney functions, 165–167
    liver functions, 162
    protein energy malnutrition, 160
    sugar impairment, 157, 158
Cerebrospinal fluid (CSF)
    clinical significance, 137
    CSF protein analysis, 135–137
    functions, 135
    occurrence, 135
    reference range, 137
    secretion, 135
Cherry Crandell Unit (CCU), 118
Cholesterol
    absorption, 93
    clinical significance, 95
    esterified cholesterol, 93
    functions, 93
    occurrence, 93
    reaction, 94
    structure, 93
Cholesterol esterase, 139, 140
Cholesterol oxidase, 139, 140, 142

Chylomicrons, 93, 139, 141–143
Clinistix, 50–51
Color reagent, 68, 69, 104
Colorimeter, 17–20, 98
Creatine, 75, 131–133
Creatine kinase (CK), 131–133, 167
Creatinine
    albumin-creatinine ratio, 33
    clinical significance, 78
    creatinine clearance, 79, 80
    metabolism, 76
    reference range, 78
    synthesis, 75
    transport, 75
Cresolphthalein, 86, 87

D
Diacetyl monoxime, 68
Diazo reagent, 98–100, 125
Dipstick, 34
Direct bilirubin, 97–101, 163
Double beam spectrophotometer, 19

E
Enzyme linked immunosorbant assay (ELISA)
    antibody, 153–155
    applications, 155
    competitive, 154
    direct, 155
    indirect, 155
    principle, 153, 154
    sandwich, 153
Equivalent weight, 22, 23
Ethylenediaminetetraacetic acid (EDTA), 6, 7, 81, 94, 107, 131, 132
External quality control, 12

F
Fibrinogen, 5, 6, 43
Flame photometer, 147–150
Fouchet's test, 123, 124, 126

G
Globulins, 33, 34, 39, 43, 48, 85, 137, 163
Glomerular filtration rate (GFR), 71, 79, 80, 92
Glomerular proteinuria, 36, 37

Index

Gluconic acid, 50, 57
Glucose oxidase (GOD), 50, 57–60, 63, 64
Glucose tolerance test (GTT) curve, 63, 64, 66
Glycerol kinase, 141
Glycerol-3-phosphate oxidase, 141
Glycosuria, 51, 53, 55, 59, 66, 157
Gmelin's test, 125
Good cholesterol, 145
Gout, 83, 160, 161
Gross error, 10, 11

H
Harrison's test, 35–36
Hazards, 1–3
Heller's test, 35
Hemoglobin buffer, 25
Hemolytic jaundice, 95, 100, 101, 109, 125, 127, 164
Henderson-Hasselbalch equation, 24, 26
Hepatic jaundice, 101, 125, 163
Hexokinase, 131, 132
High density lipoprotein (HDL), 139, 142, 143, 145, 167
Hunter's test, 125
Hypercholesterolemia, 95, 166
Hyperglycemia, 59, 159, 165
Hypoglycemia, 59, 157, 158
Hypouricemia, 83

I
Immunoglobulin, 33, 35, 37
Inorganic phosphorus, 89–92
Instrument error, 10
Internal quality control, 11
Isoelectric point (pI), 34
Isoenzymes, 107, 129, 131

K
Ketone bodies
  clinical significance, 122
  concentration, 122
  detection, 122
  ketogenesis, 119
  ketonemia, 122
  ketostix, 121
  synthesis, 119, 122
  transport, 119
  types, 119, 122
  utilization, 122

L
Lactate dehydrogenase
  activity, 129, 130
  clinical importance, 130
  distribution, 129
  isoenzymes, 129, 130
  subunits, 129
Lactose, 49–51, 57
Levey-Jenning chart, 11
Lipase
  clinical significance, 118
  mode of action, 117
  normal value, 118
  secretion, 117
Lipid phosphorus, 89
Lipid profile, 139–145, 167
Lipids, 5, 89, 139
Low density lipoprotein (LDL), 139, 142–145, 167

M
Maximum urea clearance, 71, 72
Molar solution, 21, 22
Molarity, 21, 22
Myocardial infarction, 106, 130, 131, 133, 167

N
Nephrotic syndrome, 37, 47, 48, 88, 95, 166
Nonreducing sugar, 49
Normal solution, 22, 23
Normality (N), 23

O
Obstructive jaundice, 95, 101, 125, 127, 162, 163
Oliguria, 30
Organic phosphorus, 89
Ortho-cresolphthalein complex, 86
Overflow proteinuria, 36
Overload proteinuria, 36
Oxidoreductase, 129

P

Peroxidase, 97, 139–141, 153
Phosphate buffer, 24–26, 58, 64, 65, 104, 105, 118, 130, 140
Phosphocreatine, 75, 132
Phospholipid, 93, 139, 142
Phosphorus
    clinical importance, 92
    distribution, 89
    functions, 90
    normal value, 92
    types, 89
Phosphotungstate, 32, 81
Photometry, 17–19
Physiological buffer, 25
Plasma, 5–7, 25, 33, 35, 43, 44, 47, 58, 61, 62, 64, 65, 67, 71–73, 75, 79, 81, 85, 87, 94, 107, 118, 125, 129, 131, 132, 135, 139, 141, 143, 147, 150, 165
Plasma separating tubes, 6
Pneumaturia, 39
Polyuria, 30, 158, 159
Post analytical error, 11
Post renal proteinuria, 36, 37
Postrenal uremia, 70
Potassium
    distribution, 147–150
    functions, 148
    hyperkalemia, 150
    hypokalemia, 150
    plasma concentration, 147
    significance, 149
Pre-analytical error, 10
Precision, 9, 11
Prerenal uremia, 70
Prostate-specific antigen, 112
Protein buffer, 25
Protein-free filtrate, 61, 77, 78, 82, 83, 90, 91, 95
Proteinuria, 33, 34, 36, 37, 41, 165, 166
Pyrogallol red dye, 135
Pyruvate, 103–106, 129, 130

Q

Quality control, 9–12

R

Radioimmunoassay (RIA), 151, 152
Random error, 10, 11
Reducing sugars, 49–51, 53–55, 157, 158, 166
Renal glycosuria, 66, 157
Renal uremia, 70
Reproducibility, 9, 11, 13
Rothera's test, 120–122, 158

S

Sensitivity, 9, 50, 152
Serum
    preparation, 6
Single beam spectrophotometer, 19
Sodium
    distribution, 147–150
    functions, 147
    hypernatremia, 149
    hyponatremia, 149
    plasma concentration, 147
    significance, 149
Solute, 21, 22, 24, 30
Solvent, 1, 19, 21, 139
Somogyi unit (SU), 113, 114
Specificity, 9, 17, 152, 153
Spectrophotometer, 17–20, 69, 143, 144
Standard urea clearance, 72
Stercobilin, 123
Sucrose, 49
Sulfosalicylic acid, 35, 136
Systemic error, 10

T

Transaminase, 103–106, 164
Transmittance, 17
Triglyceride, 93, 139, 141–145, 167
Tris buffer, 24, 26
Tubular proteinuria, 36, 37
Turbidimetry method, 136, 137

U

Unconjugated bilirubin, 97, 100, 101, 123, 125, 162–164
Unesterified cholesterol, 93

Urea
  blood range, 67, 70
  diagnostic importance, 32, 67–70
  excretion, 67
  synthesis, 67
  urea clearance, 71–73
  uremia, 67, 70
  urinary excretion, 70
Uric acid
  diagnostic importance, 81–84
  formation & excretion, 81, 83
  normal range (female), 83
  normal range (male), 83
  structure, 81
Urinary proteins, 33–37, 39
Urinometer, 30, 31
Uristix, 50
Urobilin, 123, 126, 127, 162

Urobilinogen
  clinical significance, 127
  formation, 123
  test (s), 126
  transport, 123
UV range, 18, 20

V

Vacutainers, 5, 6
Venous blood, 5
Very low density lipoprotein (VLDL), 139, 141–144
Visible range, 18, 20

W

Waste disposal, 1–3

Printed by Printforce, the Netherlands